VICTORY IN THE EAST
The Rise and Fall of the Imperial German Army

By

Michael P. Kihntopf

 WHITE MANE BOOKS

Copyright © 2000 by Michael P. Kihntopf

All maps were prepared by the author.

ALL RIGHTS RESERVED—No part of this book may be reproduced in any form without permission in writing from the publisher, except by a reviewer who wishes to quote brief passages in connection with a review.

This White Mane Books publication
was printed by
Beidel Printing House, Inc.
63 West Burd Street
Shippensburg, PA 17257-0152 USA

In respect for the scholarship contained herein, the acid-free paper used in this book meets the guidelines for permanence and durability of the Committee on Production Guidelines for Book Longevity of the Council on Library Resources.

For a complete list of available publications
please write
White Mane Books
Division of White Mane Publishing Company, Inc.
P.O. Box 152
Shippensburg, PA 17257-0152 USA

Library of Congress Cataloging-in-Publication Data

Kihntopf, Michael P., 1949-
    Victory in the east : the rise and fall of the Imperial German Army / by Michael P. Kihntopf.
       p. cm.
    ISBN 1-57249-148-5 (alk. paper)
    1. Germany. Heer--History--World War, 1914-1918. 2. World War, 1914-1918--Campaigns--Eastern Front. 3. Germany--History, Military. 4. Strategy--History--20th century. 5. Military art and science--Germany--History--20th century. I. Title.

D551 .K46 1999
940.54'1343--dc21

99-056333

PRINTED IN THE UNITED STATES OF AMERICA

CONTENTS

List of Illustrations iv
List of Maps v
Preface vi
Chapter One   The Transformation of an Army 1
Chapter Two   The Eastern Front 17
Chapter Three   Rumania 32
Chapter Four   A Return to the Galician Front 52
Chapter Five   Armistice 63
Appendix   Wenden, the Final Battle of the German Empire 70
Notes 78
Bibliography 90
Index 97

ILLUSTRATIONS

Composition of a Corps, 1914 3
A Section of the 3 Guard Division, Lehr Regiment, April 1916 21
Multinational Force, Germans and Turks in Galicia 28
In Remembrance of Turkish Support 56
Placement of German Divisions on the Eastern Front, November 1918 66
Comrades on the Fringe, Baltic Coast, 1919 73

MAPS

German Military Districts, 1914	4
Eastern Front, 1916	20
Brusilov Offensive, 1916	24
Rumania Surrounded by Central Powers and Their Conquests	33
Salonika Front, 1916	38
Rumanian Army's Campaign	42
Central Powers' Campaign against Rumania	45
Hermannstadt Battle, September 1916	47
Kerensky Offensive	55
Placement of German Army Units at the Time of the Armistice	64
Principal Evacuation Route on the Eastern Front	68
Battle at Wenden, 1919	76

PREFACE

The inspiration for this book arrived at my house one hot summer afternoon in 1992. Amid packing of old newspapers I found my grandfather's World War I trip books and photo album. The trip books contained place names, dates, and a rather large assortment of poems. Despite what I considered a somewhat extensive knowledge of European geography, very few of the place names were recognizable, and the poems, because of my limited German reading ability, meant even less. However, the true treasure was the photo album. Here I could see images of life during the war. Luckily, the vast majority of snapshots had dates and places written on their backs along with a few unit designations.

Painstakingly, I arranged the photos by date (they had been jumbled over the years) and considered each one. Those that had no date or place on the reverse, I positioned according to recognizable people and terrain. For a few weeks I looked at the faces that stared back mutely at me and contemplated the landscape pictures. My questions to each of the photos were simple: Who are you? Where are you? Why are you there? and What happened to you?

To many, World War I is stereotyped by incessant trench warfare that groveled over a few yards of territory from 1914 to 1918. This may be true for those who see the Western Front as the only true representation of the Great War. However, for the German soldier and his home supporters there was an equally important Eastern Front. In that theater, the Central Powers' armies were victorious to the point of actually winning all their objectives. By 1918, Austro-Hungarian, German, Bulgarian, and Turkish soldiers occupied Serbia, Montenegro, Rumania, the Baltic Coast, most of European Russia, and the Ukraine. This accomplishment is largely overlooked or down played for the benefit of the decisive victory of the *Entente* on the Western Front.

This work chronicles the events which occurred on the Eastern Front from largely a German point of view. It begins in 1916 and concludes in 1919. The beginning date came about through research which clearly showed that the German army, like all the armies in the war, had undergone a monumental

Preface

change in its make-up and application of force. While the end of 1919 was selected because fighting on the Eastern Front terminated in that year with the demobilization of principal *Freikorps* units.

This book is about the German victories on the Eastern Front during the war and immediately thereafter. Although the fighting had stopped in the West, the Baltic Coast continued to be a hot bed of war.

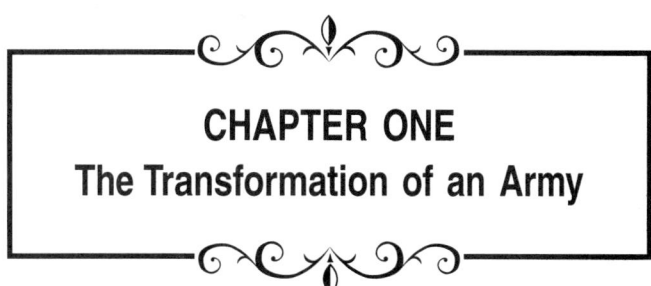

CHAPTER ONE
The Transformation of an Army

In the closing years of the nineteenth century, the German Empire found itself increasingly isolated within the European community. The French had entered into a defensive arrangement with Russia. Russia was attempting to build a commercial empire in the Balkans with links to the Serbians. While on the seas, the British Empire viewed Germany's acquisition of a battle fleet as a threat toward its power on the seas. It is little wonder that the *Oberste Heeresleitung (OHL)*, General Staff, recognized the need to plan for a two-front war. The army's organization reflected this need to be prepared. Yet, by 1916, the army was no longer what it had been in 1914. High casualty rates among its predominantly peasant soldiers commanded by aristocratic officers had led to the induction of large numbers from the middle class and from among the industrial workers. Gone too were the notions of a romantic war. Ideals of heroism had been snuffed out by the efficient application of weapons and tactics to the new battlefields.

ARMY ORGANIZATION OF 1914–1916

Militarily, Germany addressed all levels of society and state administration. The empire had been divided into 24 military districts, each responsible for raising a corps. On paper, the corps consisted of 34,000 men. Most of its officers came from a long line of professional soldiers who had ancestral estates in the military district. With few exceptions, the officers saw themselves as the last vestiges of medieval knighthood. The ranks consisted of volunteers and conscripts from the area. In effect, the corps could have been recognized as a regional army without thoughts toward the grander picture of a national army had not *OHL* installed some reminders. For instance, the XX Corps, whose headquarters was in Allenstein, Southeast Prussia, consisted of two divisions, the 41 and the 37. Each division contained soldiers from Posen, West Prussia, and the Masurian and Ermland regions almost exclusively. The four regiments in each division were identified by two names. While in barracks, the regiments of the 41 Division were the First and Fourth Posen Regiments, the First West Prussian Regiment, and the Infantry Regiment of *Deutsch Ordens*. But when OHL addressed these regiments, they became the 18, 59, 148, and the 152 Infantry Regiments.[1] The only exception to this regionalism was in the Guards Regiments which recruited from all over Prussia and the Imperial provinces of Alsace-Lorraine.

Besides the two infantry divisions, the corps had one cavalry division and one regiment of foot artillery. Each division, with supporting elements, had an authorization of 17,000 with a chain of command progressing downward through the division, brigade, regiment, battalion, company, platoon, finally ending in an eight-man section. On a peace time footing, most corps filled 65 percent of their positions. Border corps maintained an 80 percent manning level while cavalry units kept their ranks filled at 100 percent.[2] Such high manning rates meant that a corps could deploy or act as a shield while interior corps strove to reach full strength through mobilization with only minimal dependency on reserves.

Unlike other armies where officers commanded all units, supervision of the smallest elements, the platoon and section, was often the domain of non-commissioned officers in peacetime. This was largely caused by the reluctance of the aristocratic officer corps to commission those who had neither title, social standing, or popularity.[3]

Growing diplomatic tensions since the beginning of the twentieth century saw a steady increase of the army's active duty strength. In 1906, there were 562,277 on the active rolls.[4] Less than two years later, the amount rose to 609,168.[5] And by 1913, in the wake of the Moroccan crisis, there were 800,646 men in barracks or aboard ship.[6] The increase also reflected the realization by OHL that too few of the rapidly growing population were receiving military training. Germany's population had increased by over 50 percent since 1850 yet the number of corps had not substantially changed until the years before the war. The Kaiserine Army of 1914 consisted of 218 active regiments, 113 Reserve regiments, and 96 *Landwehr* regiments.[7]

Following the Clausewitz ideal of the military's non-involvement in civil administration, the corps commanders operated their domains separately from civil government during peacetime. This separation extended to the soldier. According to one participant in the 1913 army, soldiers had no social contact with civilians in their garrison towns. As a rule, the soldiers were shut up in the barrack.[8] This specifically meant that soldiers were not subject to civilian laws. The army was not only a class unto itself, it was also a state within a state. However, when war was declared, as in August 1914, this separation almost disappeared. An antiquated Prussian law went into effect by which the local government fell under the military's control. In effect, the government called a

Corps Composition, 1914

Division

Infantry Brigade		Infantry Brigade	
Infantry Regiments / Machine Gun Companies			
Field Artillery Brigade			
Field Artillery Regt.		Field Artillery Regt.	
Field Art. Group	Field Art. Group	Field Art. Group	Field Art. Group
Ammo Col.	Ammo Col.	Ammo Col.	Ammo Col.

Cavalry Regiment

Pioneer Company
Telephone Detachment
Bridging Train
Sanitary Company

Division

Infantry Brigade		Infantry Brigade	
Infantry Regiments / Machine Gun Companies			
Field Artillery Brigade			
Field Artillery Regt.		Field Artillery Regt.	
Field Art. Group	Field Art. Group	Field Art. Group	Field Art. Group
Ammo Col.	Ammo Col.	Ammo Col.	Ammo Col.

Cavalry Regiment

Pioneer Company
Telephone Detachment
Bridging Train
Sanitary Company

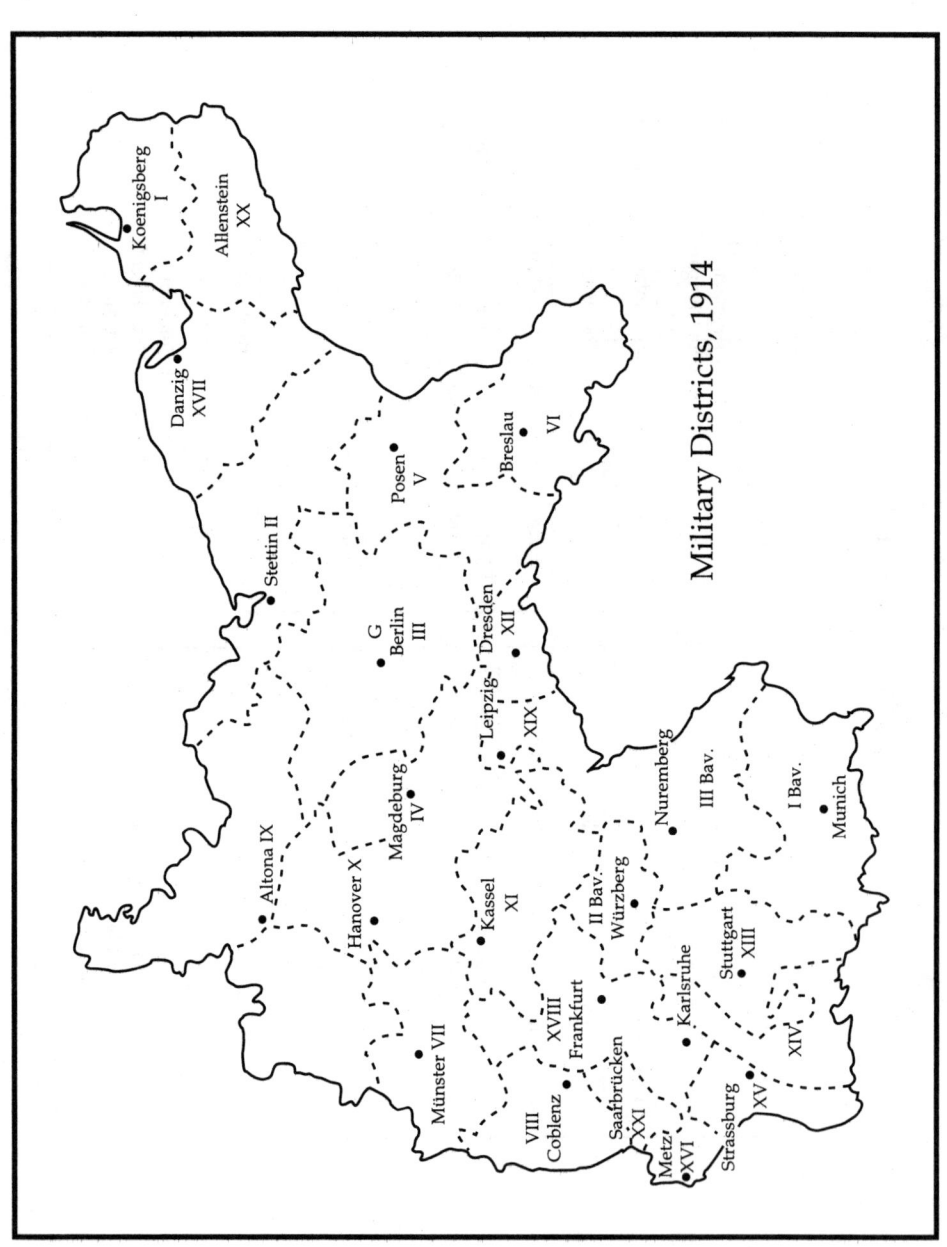

state of siege into existence.⁹ The situation was viewed as if Germany were being assailed from all sides and therefore all administration must be directed for one thing, repel the invader. Unprepared for such duties, both the military and civilian bureaucracies found that the transition to a wartime footing was anything but smooth. The change was complicated because the law redrew local jurisdictions to correspond with the boundaries of the military district.¹⁰ Quite often these new lines were not the same as the peacetime lines. The chains of authority that civilian administrators used became muddled as did the status of those civilians who were not mobilized. This confusion leaked into daily life as the war dragged on, especially in regards to the distribution of food. It is little wonder that the German civilian had no love for a military administration by 1918.

LIFE FOR THE SOLDIER, 1914–1916

Conscription had been in effect in most of the Germanys since the *levee en masse* days of the French Revolution and Napoleon. The induction of the male population was one of the ways that Germany could ensure that there were enough trained soldiers if its enemies chose to invade. Despite the fact that the empire had a highly professional volunteer force of officers and non-commissioned officers, it depended on conscription to fill the ranks.

The German male became liable for military service at 17 when he was enrolled in the *Landsturm*, an organization equivalent to a regional or national militia. No training was given nor was the youth required to report to a barracks. The *Landsturm* acted as a pool of potential soldiers should the nation need them to fill gaps in the active or Reserve regiments. However, the youth could volunteer for service at this age.¹¹ For those who chose to wait, military training began on the first of the year following their 20th birthday when one reported to the regional corps headquarters for two or three years' active service.¹² Nominally, 60 percent of those eligible for service in a year group were inducted. There were a multitude of exemptions one could apply for including schooling and family support. Exemptions were granted on a year-to-year basis. If a person was able to put off serving for three years, he was released from conscription obligation. Most exemptions were approved because of budgetary constraints imposed by the *Reichstag* on maintaining a large standing army. The army couldn't afford to outfit 100 percent of all those eligible and fit for service. Examiners disqualified some inductees because of physical or mental handicaps. But even the excluded did not escape the military machine. They were enrolled in the *Ersatz* Reserve for 12 years from which they could be called for specific duties related to supporting the active services.¹³ Nevertheless, out of a population of 67.5 million, *OHL* managed to field an army of 4.9 million in 1914.¹⁴

Those who completed active service expected to be in the Reserves for four or five years after release. Each active duty division had a counterpart in the Reserves of equal size and capability. The active duty Third Division was entirely different from the Third Reserve Division. In this way, the army retained a vast pool of organized and trained soldiers while supporting a smaller contingent on active service. Reservists met periodically for training but excuses abounded and were readily accepted by those in command. Once again budgetary constraints were a deciding factor. As a result, Reserve soldiers were not always as well clothed or equipped as the active force.

Reserve duty was followed by 11 years in the *Landwehr*, a Prussian tradition which dated back to the Napoleonic era. The *Landwehr* had been an army of non-aristocrats who had risen to the call to liberate Prussia from Napoleon during the Wars of Liberation. Traditionally, the *Reichstag* viewed these units as their countermeasure for deterring a Junker-dominated military. Throughout the nineteenth century, both the *Kaiser* and the *OHL* had tried to eliminate these units; however, popular sentiment had kept them in existence.

The completion of 11 years in the *Landwehr* brought the German male full circle back to the *Landsturm* in which he completed his military obligation at age 45. Unlike his youthful days of 17, the man in his mid-forties was a well-seasoned and -trained soldier, but he was getting on in years. *Landsturm* men spent the last days primarily as part of fortress contingents or guards for various military works.

Although the pay for a recruit in 1914 was about one and three quarters cents a day, the army supplied the soldier with everything to make life comfortable amid the rigors of training. Besides an everyday uniform, each soldier was responsible for about 70 pounds of equipment. The list was approximately the same throughout European armies.

Extra pair of pants	2 shirts	cap comforter
one pair of drawers	great coat	3 pairs of socks
cardigan	boot laces	mess tin
towel, comb, soap	toothbrush	razor and kit
latherbrush	water bottle	sewing kit
eating implements	spine protector	field dressing
tin of grease	bottle of oil	set of equipment
entrenching tool	bayonet	rifle
rifle pull through	waterproof sheet	
150 rounds of ammunition	half tent with equipment[15]	

In 1916, the gas mask was added to this list along with waterproof overalls for trench duty. The steel helmet was first introduced at Verdun in 1916 and became standard issue in 1917.[16]

In 1914, the army provided a daily food regimen of 4,038 calories to build up the men to carrying all the equipment on long marches.[17] This daily food supply consisted of 12 ounces of meat and one pound of vegetables a day. The ration was usually given to the soldier in a type of pea soup that had lumps of meat, fat, and gristle.[18] By June 1916, the meat ration had dropped to 10 ounces a day and there was one meatless day a week. That year also saw the introduction of a type of flour made from turnips. This new flour, to critics in the line, was mixed liberally with normal flour and saw dust to produce bread. The bread was almost indigestible besides being tasteless. To offset the blandness, crafty scientists perfected another application of turnips in the form of a flavored paste for spreading across the bread. On occasion, the cooks prepared a stew made of horse meat and dried vegetables which were referred to by the soldiers as barbed wire entanglements. The new cuisine was washed down by a liberal spirit ration. The soldier received approximately 1/20 of a pint of spirits and nearly a pint of wine a day.[19]

For defense, the soldier was armed with a Mauser *Gewehr* M1898 rifle which used a rimless bottleneck 7.92 millimeter cartridge fed to the breach by an internal five-round magazine. Although the rifle's range had been the nemesis of the American and British armies when the Spaniards and Boers wielded them in those respective wars, not all German soldiers were trained to use the rifle with deadly accuracy. Harking back to the days of Frederick the Great, the soldier was trained to discharge his weapon from the hip along with the rest of his element. A solid wall of unaimed musketry fire followed by a rousing bayonet charge was still preferential to striking at individuals. Sharpshooting was better left to snipers.[20]

To support the rifleman, *OHL* had adopted the Maxim 1908 machine gun. Belt fed, the Maxim was water cooled and weighed close to 123 pounds. Planners measured its effectiveness by the number of rounds it could shoot in a specified time. When accepted, the Maxim shot 600 rounds a minute. With this in mind, the machine gun was seen to equal the fire power of 60 riflemen.[21] Machine gun companies were formed as early as 1911 and assigned to the battalion level. Normally, nine machine guns were in a company. These guns could be split into three sections to support a particular, advancing company or they could remain en masse to support the battalion's advance. Their employment in battle scenarios was different than practices carried out by both England and France. In those armies the machine gun was placed in the line to augment firepower and release soldiers for added reserves. *OHL*'s concept of the machine gun was to place the weapon in reserve, thus allowing more men into the line.[22] The battle concept was still one which called for soldiers, and not technology, to advance close enough to finish the job with the bayonet.

KAISERINE TACTICS

There was very little room for tactics in the minds of pre-war planners. Almost as a whole, world leaders believed, based on the European wars of the last half of the nineteenth century, any war between industrialized nations would be very short. Economic advisers added to this belief by prophesying economic ruin to any country if a war lasted longer than a few months. They opined that no matter how rich a country might be, a modern war would require fantastic sums of money. It would also require the displacement of the working force from the factories to the battlefield, thus effectively shutting down the country's industrial base. Consequently, pre-war plans of both the *Entente* and Germany showed armies rather than divisions moving to specific objectives. The concept was that one decisive blow in the correct place would end the war. Therefore, if there were to be only one battle, then all the country's manpower must be brought to bear at the very onset of the war. The army that did not immediately assume the offensive, would be crushed.

France developed Plan XVII in which five armies were used. Initially, the First, Second, Third, and Fifth Armies would invade Germany through the Alsace-Lorraine. The Fourth Army would remain in reserve along with mobilizing troops. The only option this plan had was that, in the event Germany should violate the neutrality of either Belgium or Luxembourg, the Fifth Army would fight a delaying action along the Belgian border while the Fourth Army took its place in the invasion of Germany. French experts believed that such a large force entering Germany would call any outflanking movement in northern France to an abrupt stop and, therefore, not seriously injure the Fifth Army.

For Germany, the Schlieffen Plan, formulated in the 1890s, was the epitome of the army's strategy. It showed five armies sweeping through Belgium and Luxembourg, flanking the French army that would be drawn into action by a Sixth and Seventh Armies' fainting invasion through the Alsace-Lorraine. The French army would be encircled in a long arch which would capture Paris and sever supply lines. Then they would be caught in the grips of the five armies in its rear and the two facing it. In such a war of wide sweeping actions, tactics below the corps level were left up to the regimental commanders. Although *OHL* published operating instructions for the deployment of every echelon of commander, tradition depended on each officer making tactical decisions based on the circumstances they were under.

There were many influences on the utilization of tactics in the late nineteenth and early twentieth centuries. The first influence had been the American Civil War where German observers witnessed the devastating power of rifled bullets and entrenched fortifications. Similarly, the Boer War showed *OHL* how open battlefield maneuvering had proven superior to infantry column actions in the presence of rifled bullets and entrenchments. Tacticians saw that the distance between advancing soldiers had to be opened to an undetermined distance to allow each individual to take advantage of any terrain feature which would protect the soldier until he could safely aim and shoot individually at selected targets. Hailed as a method by which commanders could save lives, a faction within the *OHL* and at the regimental levels saw these tactics as leading to a loss of control over the advancing unit. This faction consisted of officers who had reached superior rank and had combat experience during the Franco-Prussian War of 1870–1871. What stood foremost in their repertoire of experience was the memory of the Drüheberger or squatting hares. Studies after the 1870–71 war revealed that almost one third of the soldiers involved in combat became missing as unwounded stragglers.[23] Most of these soldiers had participated in the initial attack's first volley but then either refused to move on or remained stationary and fired at anyone or anything which got near regardless of uniform. These antagonists also pointed to the recent Russo-Japanese War in which firepower had proved to be the deciding factor in the battles. Massive, determined infantry attacks had taken entrenchments.

The Field Manual of 1906 called for a company to advance in column. Two rows of soldiers, preceded by the commander, would march almost shoulder to shoulder. When the company reached a point approximately 400 to 500 yards from the enemy line, the front platoon would stop and deliver a volley of unaimed rifle fire. The second platoon would then advance slightly past the first line which was reloading and fire a second volley. These volleys were designed to keep the enemies' heads down until the company was close enough to charge into the opposing line with the bayonet. *OHL* still viewed battle as an action between individuals locked in hand-to-hand combat. Open field operation would neither deliver the necessary firepower nor insure that all would participate in the final bayonet charge. The fear of soldiers deserting from battle still permeated the officer ranks. This was seen in the positioning of the NCO behind the two columns. From this placement, he could either add a kick when the line failed to advance to the proper yardage or to stop those who attempted to leave the battle.[24]

The controversy over the type of tactics to be followed must have been a dividing factor among unit commanders. The only compromise came in the way of broadening the space between men to six meters but the concept of hand-to-hand combat was not dropped. *OHL* saw the beginning battles of 1914 in the same light as did General Friedrich von Bernhardi in his 1913 book:

> The attack must press on irresistibly; losses must not be shirked...artillery fire must accompany the infantry to the very last stage of the attack, and finally with common shell, even at the risk of inflicting losses on their own troops. The storming infantry must throw hand grenades inside the hostile position, while the heavy howitzer shells continue to burst in and behind the position.[25]

This tactic proved devastating in the first months of the war as German soldiers advanced in columns against machine guns and the well-aimed musketry of the British army. Numerous newspaper reports described how reporters were shown whole companies lying dead in formation. Yet it was not wholly as a result of large casualties that unit movements became passé.

NEW WEAPONS

The end of 1914 marked the end of the romanticism that many had carried into battle on those warm August days and nurtured through the Christmas Truce. Nineteen hundred and fifteen saw the huge armies of the *Entente* and the Central Powers settling into a war of attrition. Trenches were dug from the North Sea coast to the Alps and from the Baltic to the Rumanian border. Sweeping corps movements were no longer possible. Each trench had become a fortress to which the opposing army laid siege. With this concept in mind, opposing engineers, termed as pioneers, began to adopt weapons that could effectively give their side the wherewithal to breach the others' defenses.

The first development was the hand grenade. Perhaps one of the oldest gunpowder-using weapons known to the soldier, the first grenades were clay pots stuffed full of gunpowder and other debris, then sealed at the top except for a protruding fuse. A brave soldier then lit the fuse and threw the pot down among besieging forces. Depending on the type of debris in the pot, the weapon had no effect beyond that of a concussion designed to startle more than injure. Both sides improvised in the early days of trench warfare. Both the *Entente* and Central Powers forces used a primitive device called a hairbrush in 1914 and early 1915. The hairbrush was a piece of wood cut in the shape of a handled brush to which engineers strapped bulk explosives. A thrower lit an ignition fuse, waited, and then threw it into the enemy's position. The British followed this apparatus with an equally primitive weapon called the jam-tin bomb which was essentially the hairbrush explosives stuffed into discarded food tins. In either case, because of the lack of projecting debris, the first grenades did little more than cause duress through loud explosions.[26] However, *OHL* recognized the capabilities of such weapons. German industry soon developed a number of effective products.

The tortoiseshell was the first to appear in 1915. It had a flat circular shape with percussion detonators all around its outer edge. Designed to explode on impact, the tortoiseshell proved to be unreliable in mud or as an easily transportable weapon. Another failing was that it still only produced a big bang rather than casualties. Along with the tortoiseshell there appeared a grenade which became as much a symbol of the German soldier as his helmet. Reminiscent of a potato masher, it was a nine-inch by two-inch cylinder filled with explosives and shrapnel on the end of a five-inch stick to which streamers were often attached to stabilize flight. Rather than rely on impact, the potato masher had a timed fuse that was initiated when the soldier pulled down on a string in the handle. This caused a spring to puncture a cap and light a fuse. The reliance on a timed fuse rather than impact to detonate the charge made the new grenade reliable in any kind of weather or terrain as well as transportable. But its most resilient trait was that the shrapnel in the grenade actually produced casualties along with a demoralizing explosion. Improvements to the detonation system in later models led to its preferred use by the soldier as the weapon of first contact with the enemy.

The second weapon which destroyed the romanticism of battle was the employment of poisonous gas. Poisonous gas was another ancient weapon. Its use was first chronicled in the Peloponnesian War of 431–404 BCE when besieging Spartans soaked wood in pitch and sulfur, then burned them beside cities' walls in an effort to incapacitate the defenders. American Union generals had turned down a proposed idea of using chlorine gas on entrenched Confederate soldiers.[27] The hideousness of this weapon was recognized at various Hague Conventions. Both the 1899 and 1906 Conventions prohibited its use as inhumane.[28] But by 1915, ideals were far removed from those of the Conventions. German leaders saw a need to maximize their power to inflict casualties without sacrificing any more of their own soldiers.

The first German application of gas occurred on the Eastern Front in January 1915, during the advance on Warsaw.[29] Pioneers moved canisters of xylyl bromide close to the Russian trenches and opened the valves. The effect of the tear-inducing gas was far from effective because of the cold temperatures which prevented the gas from evaporating. However, the psychological effect was most devastating. Russian soldiers abandoned their positions, allowing a temporary advance, but the slow dissipation also prevented the Germans from consolidating their gains. Scientists sent specifically to observe the gas attack took note of the conditions and went back to the laboratory to optimize gas assault plans.

The second deployment of gas came on the Ypres front on Pilchen Ridge in April 1915.[30] Once again, pioneers risked their lives to place 5,730 cylinders of chlorine gas within a few feet of the French trenches. On the morning of the twenty-second, after a one-hour barrage, the cylinder's valves were opened. Within a few minutes, a yellow-green cloud formed that was estimated to be 300 to 400 yards wide and a half mile deep. The cloud was carried by a mild five-knot wind across the remaining yards of no man's land and into the positions of the 45 Algerian Division and the 87 Territorial Division. Panic ensued and the French abandoned their trenches. At day's end the Germans had rounded up 2,000 prisoners and 51 guns. German losses were insignificant according to the battle report.

War has a way of bringing out the innovativeness in man. Some factory workers noticed that there was quite a bit of space in an artillery projectile which went unused. In their opinion, various gases could be interlaced with the explosive charge of the projectile. Such a delivery system was probably preferential to the pioneers who suffered casualties in deploying canisters as well as during the gas release should there be a sudden shift in the wind direction. Trial shells were used on the Western Front. The method proved less than effective since the explosion, needed to release the gas from the shell's inner casing, often dissipated the gas before it could do any harm. Additionally, the small amount of gas that could be placed in the shells wasn't of any practical quantity. Too many shells would have to be fired to have any appreciable effect. Although these problems were later solved through the invention of a shell which delivered gas instead of high explosives, the Germans continued to rely on canisters to deploy gases. Artillery gas shells proved to be effective as counterbattery ammunition. Whereas high explosives may cripple a cannon or two, gas shells, in sufficient quantity, could saturate an area thought to contain the enemy's field artillery. The guns would become inoperable or, at the least, not as accurate or rapid in their fire since aimers and loaders would be hampered by protective equipment.

Gases were classified into two categories, lethal and irritants.[31] Irritants could be just as deadly as those gases considered lethal. Lethal gases were designed to kill upon contact or inhalation. The base of these gases ranged from cyanide to arsenic. They were classified as blue cross because of the color of the cross, painted on the canisters or shells, to indicate the type of gas inside. Although used in varying quantities, lethal gases were not as prevalent as irritants. Irritant gases, termed green and yellow cross, consisted of all those which caused discomfort to the enemy. Initially, this could be as simple as tear gas or the itching powder many students had used in school; however, irritants also caused prolonged suffering as in the case of phosgene and chlorine. Both of these gases caused severe irritation to the lungs or eyes. In many instances, the victim had no idea that he was gassed. Sometimes 48 hours would elapse before painful symptoms occurred. Lung membranes would blister and the blisters would eventually rupture during coughing, causing the soldier to drown in his own fluids. Mustard gas was the preferred type from the irritant series because of its incapacitating capabilities and its longevity in the atmosphere. It sometimes lasted up to several days, occupying lower levels such as shell holes or dugouts.

Far from being an effective weapon, the use of gas appears to have been primarily for the purpose of weighing down or incapacitating the enemy rather than to kill him. Although the official casualty lists do not contain deaths caused by gas in the trenches or in no man's land, throughout the war, hospitals recorded that 91,198 soldiers died from gas-related injuries on all fronts. Nonfatal injuries totaled 1,205,655 of which Russian soldiers suffered the most. They accounted for over one third of the nonfatal casualties and more than 50 percent of the deaths.[32]

The final weapon which ended the *joie de vivre* of battle was the flamethrower. Fire had always been a member of the notorious arsenal of siege weapons. As far back as history cares to recall, defenders pouring flaming oil on besiegers and attackers often "smoked" the enemy out into the open for the final fight. The most effective fire weapon was referred to as "Greek Fire." Deployed by the Byzantines aboard their ships in the sixth century, the incendiary device was instrumental in establishing Byzantine sea power. Some historians have even attributed the Eastern Roman Empire's long life to this device.

Greek Fire could be deployed as either a flamethrowing apparatus or as "shells."[33] The shell was loaded onto a catapult, lit, and released in the direction of the enemy ship. When the projectile splattered on the deck of the ship, the oil quickly ignited the oils already impregnated in the ship's planking. The fire was almost impossible to extinguish. The flamethrowing apparatus was normally a caldron fitted with intake and output tubing. Inside the pot was a flaming mass of oil. Air was forced into the intake tube which in turn forced the flaming oils through the output tube. The exact mixture of the oils was a state secret which died with its inventor.

German weapons developers began experimenting with various types of flamethrower weapons before the war. By mid-1911, engineers had two working models. One was a large monstrosity, about five feet tall, which required a number of men to operate valves and a long hose. The other was a smaller variety that one person could carry while another aimed a hose toward the intended target. Both models operated identically in that a steady stream of nitrogen propelled a jet of flaming oil to a range of 40 meters for one minute. Death was inevitable for those caught in the fire stream but, like gas, the weapon's real potential was in the fear that it induced. Many soldiers viewed bullets and artillery as unseen objects. Not being able to see these tools of death made some feel braver. However, fire was a very real and visible danger. Courageous soldiers ran when confronted with a billowing cloud that consumed the very air. As *OHL* had recognized the merit of massing machine guns into companies, so too did they recognize the potential of concentrating flamethrowers. In 1912, the 3 Guard Pioneer Regiment was formed around 12 flamethrowers.[34]

The static trench warfare on the Western Front of early 1915 was an ideal setting for testing the effectiveness of the flamethrower. In February 1915, the weapons were used against the French around Verdun. Then, again, in July of the same year the weapon was used against the British at Hooge Chateau. In both instances, the weapon was successful in producing mass confusion among the defenders. Casualties due to the flamethrower were minimal since the flame jet was easily avoided.[35] What caused the most damage was the infantry that followed closely behind the jets. Momentarily numbed by the appearance of a large fire ball, both the French and British were easy marks for the bayonet or a well-placed grenade.

THE NEW ARMY OF 1916

Even though the entrenched battlefield on both the Eastern and Western Fronts of 1915 may have deterred large, sweeping corps movements, the real demise of such tactics came as a result of a change in players. Gone were the aristocratic officers, idealistic students, and peasant soldiers of 1914. More than half of them had succumbed to enemy fire.[36] The officer ranks were reluctantly thrown open to professional men from the middle class, people who had managed businesses, while the ranks were filled with newly mobilized urban workers salted liberally with rural laborers. These new combatants grasped the importance of a decisive win and modeled their strategy and tactics to fit the new environment.

Strategically, *OHL* perceived two problems by 1916. The first, a reorganization of the army's structure, was well within their purview to control. However, the second dilemma, propping up the failing Austro-Hungarian armies, was not so easily remedied. Yet the solution to the first problem helped to treat the symptoms of the second as well as change the tactical battlefield for the next three years.

The management of trench warfare was a business founded in routine. Soldiers were used on three levels. When a division was put into the line, normally one third of its strength went in the front trenches, a second third in reserve directly behind the front line, and the final third at "rest." The best way to divide these echelons was by regiment, but since most divisions had four regiments, one regiment was too often split up over all three levels. In 1915, *OHL* decided that a three-regiment division was preferential and so reduced the divisions.[37] But instead of melding the extra regiment into the other three and continuing the concept of regionalism among corps, *OHL* reorganized them into new divisions or brigades and eventually into new armies. The purpose of these new units was not to defend a particular region in the *Reich* as tradition would have it. Instead, they were formed for specific purposes. An example of this structuring of forces for an objective was in the forming of the 10 Infantry Brigade in January 1915. Consisting of six battalions and one regiment of *Landwehr* troops plus two field artillery regiments, the brigade's function was to capture the Vregny Plateau near Soissons.

The attack on Vregny Plateau began with a barrage which concentrated on the front trenches and machine gun emplacements.[38] Directly behind the barrage came a group of pioneers who advanced as individuals rather than in column to cut the wire in front of the French trenches. To keep the enemy who had survived the barrage from shooting the cutters, other pioneers tossed grenades. Once the wire was cut sufficiently, the infantry advanced through the gaps with bayonets fixed. The result was that the plateau was taken with less casualties than a frontal attack normally caused. The success of this new tactic spurred further testing with a new unit called the Special Assault Unit.

Command of the new unit was given to Captain Ulrich Rohr of the Guard Rifle Battalion. Drawing from his own personal experience, Rohr put together a self-sufficient unit which would precede the bulk of infantry but not in the simple manner in which the 10 Brigade had. The Special Assault Unit had much more firepower. It consisted of a company of pioneers, a machine gun platoon (two model 1908 Maxims), a trench mortar platoon (four light mortars), and a platoon of flamethrowers (six small units). To this contingent, he added six field cannons[39] which provided close artillery support, thus freeing the Assault Unit from reliance on indirect fire from batteries at the regimental level. By mid-1915, the unit was ready for a test in the Vosges Mountains.

As with other units, training for the assault included an attack on a model of the enemy's trenches constructed in the rear area far from prying eyes. But there the similarities ended. Instead of only the officers knowing what the objective would be, each member received a thorough briefing on the unit's attack goal. This was done for two purposes. First, the attacking soldier would advance to the objective under open battlefield direction instead of column. In short, he was on his own to arrive at the right place at the right time. As such, he had to be as familiar with the mission as he would be with the terrain. The second reason was even more radical than open field tactics. Rohr saw that leadership should be placed in the hands of a middle manager rather than the upper level supervision. This "foreman" would be an NCO. NCOs had been those who initially trained platoons and elements so they could respond to officers' commands. It was only natural that NCOs would lead smaller elements in battle to ensure their tactical training paid off.

The assault was opened by six flamethrowers which saturated a section of trench for approximately one minute.[40] The Assault Unit's machine guns, trench mortars, and field battery immediately opened fire to suppress the French machine guns and field artillery. Behind the flamethrowers came two squads armed with grenades, sandbags, and wire cutters. They had worked their way through

The Transformation of an Army

the wire entanglements before the assault began to get within a few feet of the French trench. Each squad entered the front line trench on a flank and quickly established a safe zone by wedging sandbags in the trench's open ends. Using grenades, some of the squad members defended the makeshift sandbag barricade from French reinforcements in the same trench while the rest fought to close the gap between them and the other flanking squad. Using grenades and pistols, the two squads eliminated any resistance. Positions which were too strong for the bombers were isolated and Rohr's trench cannon was brought up to eliminate them at point blank range. Once the trench breach was secure, the supporting infantry came up to replace the pioneers and advance through the breach they had created in the trench. These new troops were fresh and had suffered little or no casualties moving into the breach, consequently when the inevitable French counterattack came it met strong resistance. The breach was maintained and the test was a success.

Over the next months, these tactics were improved upon. Rather than stopping at the first trench, the Assault Unit continued its advance across all trenches, bypassing strong points and allowing the trench cannon or mortar to deal with them. The result was a much larger gap in the trench into which fresh infantry flowed, eventually reaching the rear echelons. The only vestige of former tactics Rohr kept as part of the assault was a reliance on the box barrage to keep out reinforcements and supplies from the intended attack area. Artillery fire was rarely used to cut wire as before.

Reports of the new tactics soon spread. Copies of the Assault Unit began appearing everywhere under various names such as *Lehrabteilung*, *Sturmtruppe*, *Jagdkommando*, or *Stosstruppen*. And with the name came a new look to the soldier. Jackboots were replaced by half boots with puttees. Carbines superseded rifles, and harnesses for carrying ammunition gave way to canvas bags that carried grenades. Wherever possible, the *Stosstruppen* were armed with automatic pistols and other short-range weapons. The bayonet had been replaced.

Successful tactics led to heightened morale among those designated to the new unit. Open battlefield tactics called for a special reliance or camaraderie between members in the same unit. Arriving at the combat objective was an individual responsibility, but once there, the unit had to perform as a team. Soldiers formed a special trust in each other to be in the right place at the right time. The fear that pre-1914 army planners had about individuals shirking their duties if not under the constant watchful eye of an officer very seldom materialized in the new units. Those who displayed such tendencies were returned to "common" infantry duties. The high morale was also encouraged by additional rations and motorized transportation to the unit's jumping off point for an assault. Many soldiers became incapable of leaving the front for extended periods of time as did Helmut Zschulte:

> I am restless. I hate the kitchen table at which I am writing. I lost patience over a book. I should like to push the landscape aside as it irritates me. I must get to the front. I must again hear the shells roaring up into the sky and the

desolate valley echoing the sound. I must get back to my company....live once more in the realm of death.⁴¹

A leaner and more efficient German army came just in time as far as *OHL* was concerned. It allowed for the treatment of the symptoms of their other strategic problem, ensuring the Austro-Hungarian army didn't collapse. As early as November 1914, *OHL* had been concerned with Austria-Hungary's capability to pursue the war. General Erich von Falkenhayn, the war minister and *OHL* leader, had created the Ninth, Tenth, and Eleventh Armies in response to Austria-Hungary's call for assistance in fighting the Russians who had been able to occupy most of the empire's Galician province. *OHL* had even gone so far as to create the *Südarmee*, a mixture of Austrian and German divisions to bolster the Galician Front.⁴² The *Südarmee* was not only a mainstay at the front, it was also an intended training cadre for its ally's regiments. In his diary, General Max Hoffman, a member of Field Marshal Paul von Hindenburg's staff from 1914 through 1916, included numerous entries about sending divisions to prop up the Austro-Hungarian forces. From his point of view, the Austrians had been defeated within a few months of the war's beginning.⁴³ It was up to Germany to keep the sagging empire afloat.

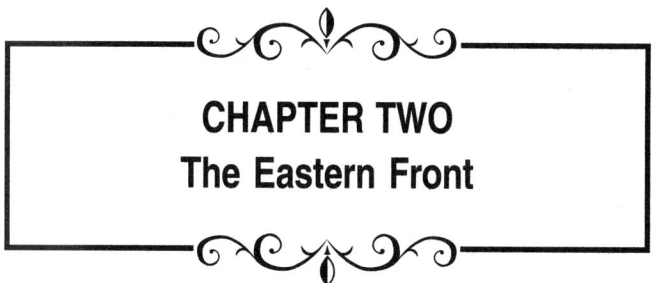

CHAPTER TWO
The Eastern Front

The change in German tactics by 1916 fit neatly into the political strategy for the Eastern Front envisioned by *Reich*'s Chancellor Theobald von Bethmann-Hollweg in 1914. In September of that year he stated that the empire's goal was that "Russia must be thrust back as far as possible from Germany's eastern borders and her domination over the non-Russian vassal peoples broken."[1] Despite what appeared to be a role for the army to fulfill, pushing Russia back was to be done more through political maneuvers than by brute force.

Bethmann-Hollweg intended to carry out the political capitulation of Russia through fomenting internal disorder by supporting regional non-Russians' bids for independence. Those areas of interest included the seaports of the Baltic Provinces, the black soil region of the Ukraine, and the mineral and oil rich Caucasus Region. In effect, he would rip the most vital areas from Russia, depriving it of over half its population and farming areas. Rather than just crush Russia's army and fighting spirit, Bethmann-Hollweg intended to deter Russia from having the means to field its vast army in the current war and ever after. He also insured that there would be no thoughts of revenge against Germany as there had been when the empire annexed Alsace-Lorraine from France. The areas carved out of Russia would be independent, sovereign nations tied to Germany only by commercial treaties. If Russia intended revenge or future war, the fighting would occur on the soil of those nations first rather than within the empire.

As early as September of the first year of the war, the chancellery had founded associations, termed as leagues, for a particular nationalistic group within the most important provinces. Count Bernhard von der Schulenburg led the Georgian League in Turkey, while the League for the Liberation of the Ukraine met in Berlin.[2] The following year, Germans residing on the Baltic coast began lobbying in Berlin for support of an arrangement which would put Courland and Livonia under German suzerainty. The Georgian League gathered soldiers from the Caucasian provinces for a contingent which, under the command of German officers, would fight with the Turks. However, Turkish political aspirations in the same area quickly led to disagreements and the League was dissolved by 1915. The Baltic and Ukrainian Leagues fared better but not necessarily along German political desires. The Ukrainian representatives were exiled socialists who wanted land reform rather than the return of the old Hetmans and grand princes[3] as the chancellery envisioned as the future for their area. The Baltic Germans on the other hand were very compliant with Berlin's desires. They envisioned no real changes to the social fabric of their provinces whether the other 96 percent of the area's population, the Letts, Estonians, and Lithuanians, approved or not.[4]

Commercially, the huge buffer zone that the new independent nations would establish between the empire and Russia would effectively exclude Russia from any future Balkan commercial developments. Middle and Eastern Europe would be closed to Russian influences and open for Germanic development. Consequently, any military objectives would have to be limited in nature so as to ensure that economic assets were not too badly damaged that they couldn't be rebuilt relatively cheaply. The army would provide a boost only where needed by a people exercising their right for self-determination. This concept, to a varying degree, was reflected in the political and military desires of the Austro-Hungarian Empire.

The Dual Monarchy's political and military objectives were to subjugate Serbia and weaken Russia lastingly.[5] Unlike Germany, who saw its main foe in the West, Austria-Hungary saw its enemy in the South. Russia could be fended off with Germany's help until Serbia fell. Once Serbia had capitulated, Russia would have no other option but to come to a negotiated peace. But the German *OHL*, considering that its prewar plans called for its major effort to be against France, was not in a position to effectively shield the Habsburg realm against the Tsarist armies. In fact, *OHL* saw the one disability of its prewar plan as being an invasion of East Prussia before France was defeated. It was willing to sacrifice that area for a limited time so that it could effectively deal with France.

The Schlieffen Plan of the 1890s placed the defeat of France as the main objective of a war with the Dual Alliance of that country and Russia. Unlike his predecessors, who had seen Russia as the most viable threat to the empire, Field Marshal Alfred von Schlieffen perceived France as the stronger of the alliance. He reasoned that Russia, hampered by its vast distances and inadequate internal communications and transportation, would have trouble completing a speedy mobilization. Schlieffen, therefore, switched the army's offensive role from the East to the West.[6] The concept was to force a quick capitulation of France and then turn the army east to link up with its Austro-Hungarian ally to meet the slowly advancing Russians. Until the French were defeated, a small contingent[7] would defend Eastern Prussia and Posen from what would probably have been an equally weak Russian incursion while Austro-Hungarian forces would divert attention with an invasion of Russia through Galicia. This concept was reinforced in the 1913 Law which appropriated money for the improvement of fortifications in the east instead of increasing the number of corps or soldiers on active duty as had the Laws of 1911 and 1912.[8] With the Dual Monarchy concerned with Serbia, *OHL*'s decision to go west effectively left the entire eastern borders of both Empires weakly held and open to Russian invasion.

The time of 1914 and early 1915 was a period in which both empires dedicated their armies to recovering territory occupied by the Russians. By mid-1915, Germany had recovered the majority of its eastern province and began a series of offensives with limited objectives designed to tie down Russian forces until France was defeated.

EARLY SUCCESSES OF THE CENTRAL POWERS

German commanders had met with varied successes in the East in 1914 and 1915. Initially shocked by the Russian army's quick invasion of East Prussia, Germany's Eighth Army, under the dual command of Generals Paul von Beneckendorf und von Hindenburg and Erich Ludendorff, had stopped the Russian advance at the Battles of Tannenberg and the Masurian Lakes, inflicting enormous casualties on the Russians in people and material. At Tannenberg alone the Tsar's army lost 92,000 soldiers as prisoners, 30,000 as wounded, and 70,000 as dead.[9] Yet Russia was able to respond with its usual tenacity. Mobilization caught up with the advancing armies, allowing the newly filled ranks to meet, defeat, and hold German forces before Warsaw for the rest of 1914 and part of 1915.

Austria-Hungary, on the other hand, was far less successful in meeting the Slavic armies. The Russian Fourth, Fifth, Third, and Eighth Armies had forced their way into Galicia in the first few weeks of the war. By September 26, they had reached the Carpathian passes in the south, threatening the Hungarian Plain, while in the north Cracow was imperiled. By far the worst outcome of the first few months of war was the casualties the empire's army incurred. Out of 900,000 soldiers the Austrians had initially deployed against the Russians, 250,000 were dead and 100,000 were prisoners.[10] The Russian advance toward Cracow and eventually German Silesia was narrowly averted by the quick formation and deployment of the German Ninth Army, one of many armies formed by *OHL* to bolster the failing Austro-Hungarian forces.

By May 1915, Germany had created the Ninth, Tenth, and Eleventh Armies specifically to augment the Dual Monarchy's forces. In a show of solidarity, *OHL* had cooperated with the Austrian chief of staff, Field Marshal Franz Conrad von Hotzendorf, to create the *Südarmee* which took up position between the Austrian Second and Seventh Armies on the extreme right of the Galician front. The Südarmee consisted of one German infantry division and nine Austro-Hungarian divisions.[11] Placed in the line to strengthen its ally, the German division also provided much needed training in the ways of modern warfare.

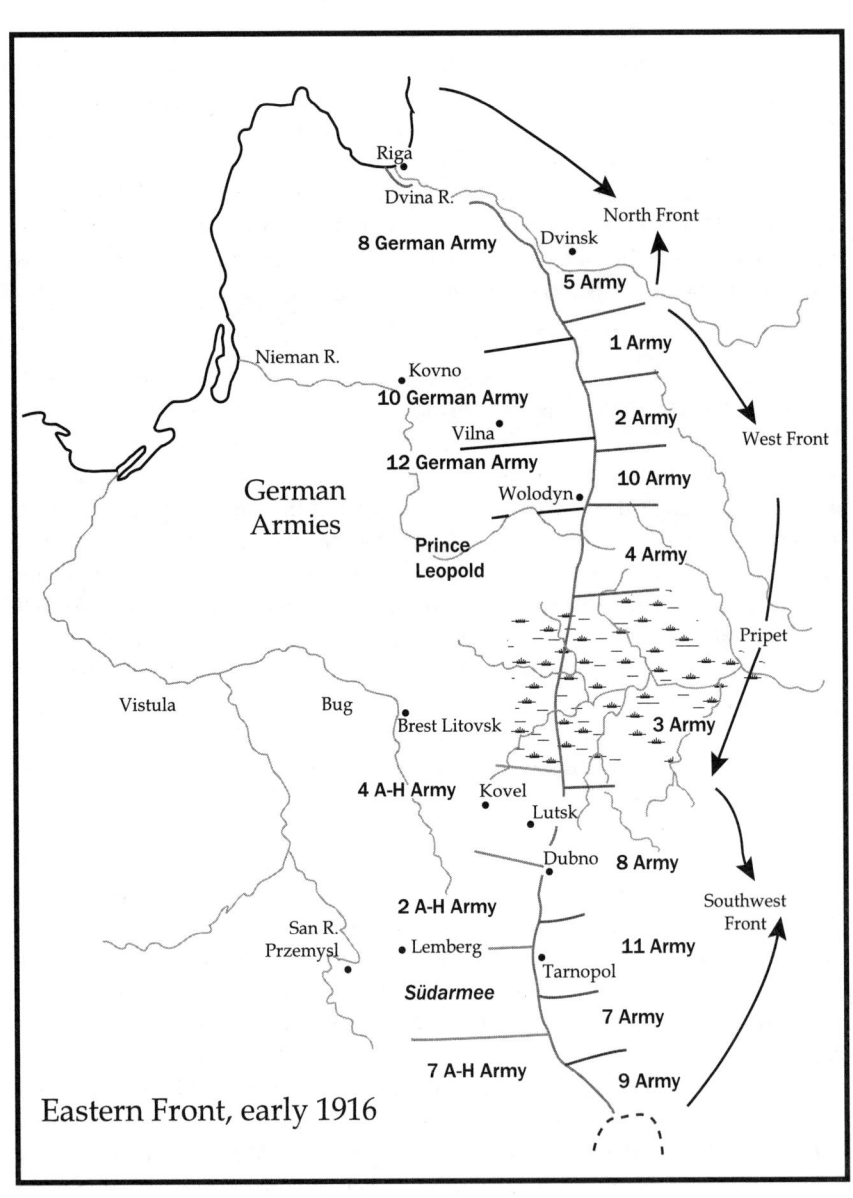

General Max Hoffman,[12] perhaps the most gifted tactician on Ludendorff's staff, noted the amount of German soldiers being sent to the Austrians throughout 1915.[13] In his diary entry for March 6, he put forth the opinion that Austria-Hungary was already a defeated nation.[14] This sentiment permeated *OHL* thought about the Eastern Front. The Dual Monarchy's soldiers were not the assets *OHL* had supposed. Consequently, German soldiers provided the majority of offensive actions as well as the steel of any defense. They would have to hold Russian armies by a series of limited victories that had significant effects. To the Habsburg Empire, limited objective warfare was a slow death. It had started the war dependent on German industry to supply its war material despite its massive Skoda works. With the Russians occupying its fertile Galician province, by mid-1915, the empire began importing grain from Germany.[15] However, in that same year two items came to the Central Powers' aid in their struggle with Russia. One was in the form of developing new tactics and the other was in Russian corruption.

General August von Mackensen was a cavalry officer. Cavalry officers are used to a war of movement with grand sweeping raids far to the rear of the enemy. Von Mackensen saw the war in the East in that light but, realizing that limited manpower and material resources were available, he intended to wage a new kind of war. Ably aided by his chief of operations, General Hans von Seeckt, von Mackensen looked upon trenches as fortifications. In all wars, fortifications

A section of the 3 Guard Division, Lehr Regiment, *Südarmee* Galicia, April 3, 1916

Author's Collection

were breached by bringing artillery firepower to bear. Anticipating the methods of the French General Henri Pétain in the next year, von Mackensen refused to use troops against technology. Rather, he would use technology against the technology of fortifications, thus saving lives.

OHL had formed von Mackensen's Eleventh Army in March and April of 1915 by moving surplus brigades from the West.[16] Its concentration on the Polish-Galician Front was hidden by having the soldiers wear Austro-Hungarian uniforms. Organized just behind the Austro-Hungarian Third and Fourth Armies, the Eleventh, along with the Third, was to drive into the flank of the Russian Fourth Army by way of a frontal attack on the Russian Third Army. This was to force a withdrawal of the Russian Second Army from the defense of Warsaw. To accomplish that task, the Eleventh was given a uniqueness in artillery. Its artillery included an estimated 302 field and 146 heavy guns.[17] Coupled with the Austrian Fourth Army's artillery of 305 field and 103 heavy cannons,[18] this was the largest concentration of fire power yet seen in the war. This allowed for one gun for every 22 meters of front with every fifth gun one of a heavy caliber.[19]

The artillery of the two armies opened fire on the morning of May 1 along a narrow front from Tarnow to Gorlice, an area approximately 25 miles long,[20] with the intention of calibrating their guns. As the day wore on, the shelling became more intense. Finally, at 0600, May 2, the bombardment reached a fever pitch and continued for four hours. The guns fired an average of 1,000 shells a minute.[21] The Russians, having the majority of their men in the front line trenches, were decimated. Despite the fact that the units had been in place over the winter, they had not constructed more than three lines of trenches. Nor were artillery batteries effectively screened or sheltered. Following the barrage came assault detachments. Using light machine guns and grenades, the detachments overcame the weak resistance of the front trenches and created "shoulders" of defense in the existing works. In effect, they created a hole or corridor in the Russian lines. Once established, the corridor was widened by linking the various assault detachments together. Into the gap were thrown fresh masses of infantry which spread out in the rear of the Russian trenches.

The Russian Third Army fell back not as an ordered unit but in a rout. By May 15, the Eleventh Army had advanced more than 60 miles, creating a salient which exposed the flank of the Russian Fourth and Eighth Armies, forcing them to withdraw. This, in turn, exposed the flank of the next army to the north and the south, causing further withdrawals. By August 15, the Eleventh and the Austrian Fourth and Third Armies had advanced 150 miles. Warsaw was taken by the German Ninth Army and Lemberg by the Austrian Second Army in the north and south respectively. The advance continued until September 30 when supply lines became too lengthy and the soldiers too tired to continue.[22] In five months, the Central Powers had advanced more than 200 miles into Russian territory. The offensive had redrawn the front from an inwardly sagging line extending from Memel in the north to the Rumanian border in the south to an almost straight line from just outside Riga to the same point on the Rumanian frontier. The center was positioned on Novogrodek, Pinsk, and Tarnopol.

The Russian army's failure to stop the German advance has been attributed to poor generalship and superior planning on the part of the Central Powers. It was true that the Third Army's commander had failed to take von Mackensen's attack seriously, even after the initial breakthroughs, and had not called for reinforcements; however, indications are that soldiers from other areas could not have stopped the drive even if German objectives had been known ahead of time. The real reason for the defeat was that the soldiers lacked the ammunition to meet firepower with firepower.

Although the German army had entered the war as the best prepared, Russia had entered with a far greater amount of soldiers. The Tsar's army before August 1914 had a strength of 1,800,000 rifles to the Germans' 621,000.[23] Yet the masses were unequipped and remained so for more than a year. Estimates showed that reserves of rifles were short by one million and there was no war industry to quickly remedy the shortage.[24] In the first seven months of the war, Russian industry produced only 41 rifles.[25] The same shortage was noted in artillery. Between 1900 and 1914, factories had produced 1,237 field guns each year and no heavy cannons.[26] For these pieces approximately seven million rounds were available. To replace the spent rounds at Tannenberg, the Masurian Lakes, or in Galicia, industry provided a mere 650,000 shells and 1,500,000 fuses throughout 1914.[27] During von Mackensen's initial bombardment in May 1915, individual Russian guns were limited to firing only a few round per day, disallowing any appreciable counterbattery fire. Most guns had only 40 rounds available while howitzers were limited to firing two rounds a day. For the heavy artillery, production didn't begin until early 1916. Machine gun production was equally stunted. Only 1,100 were produced in 1914.[28] This scarcity of material was compounded by the lack of railroads. The entire Eastern Front from Lithuania to the Rumanian border had no more than five main rail lines from the Russian interior. Close to the front these lines branched into very few side tracks. Not only did this limit the resupply of the front lines, it also restricted the movement of reinforcements and stifled any quick withdrawal to better positions.

Despite the lack of material, the Russian army was able to hold a number of lines after von Mackensen's initial gains. An English nurse who accompanied the Russians on their withdrawal noted many times in her diary that the army was pulling back to newly prepared fortifications.[29] When this happened, von Mackensen brought up his artillery and pounded a new hole in the line. Without counterbattery fire, the Russians could do nothing more than pull back to another and another entrenched position. The three major stands centered on natural defenses such as the San River on June 1, the Bug River on August 15, and the Pripet Marshes on September 30. At those locations, the expanse of water and small arms fire were enough to momentarily stop the advance. Luckily for the Russians, the Germans, with supply lines overextended by September 30,

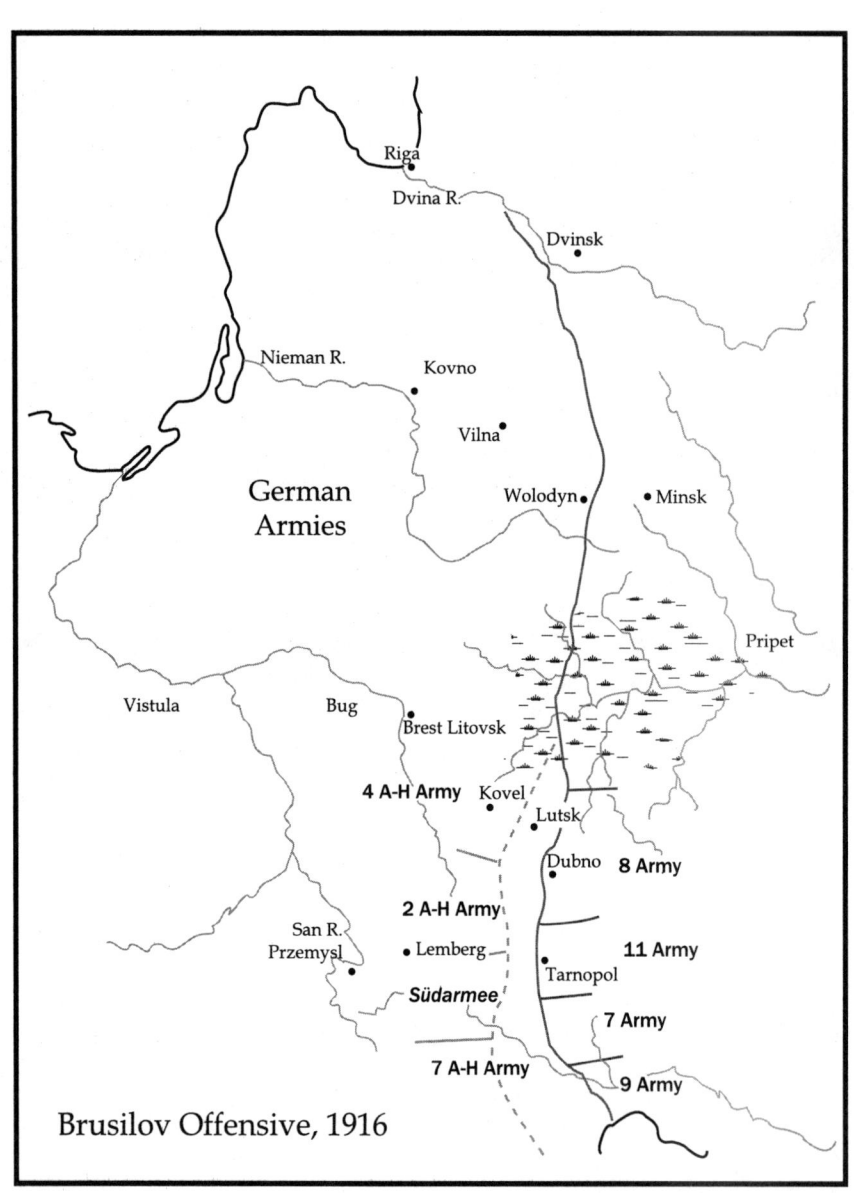

had met their limited objective of stunning the Russian army and clearing East Prussia. The Central Powers could turn their attention to other fronts. Germany looked to Verdun while Austria-Hungary saw an opportunity to punish the Italians who had just entered the war on the side of the *Entente*. This respite gave the Russian general staff and industries a much needed breather.[30]

THE BRUSILOV OFFENSIVE

Throughout 1914 and 1915, the Russian general staff, Stavka, had tried various tactics to break the German or Austro-Hungarian trench lines. In the beginning, they had used the time-honored tactic of en masse advance with or without a preparatory artillery barrage just as all the other European armies had used. As in the west, defensive machine guns and massed artillery fire had taken their toll on the Russian soldier. Late in 1915, the Russians attempted to emulate von Mackensen's tactics of a short, massive artillery barrage followed by an assault on a very narrow front. These plans too only led to disaster. Ill-supplied with shells for the most part of 1915, artillery officers could not afford the expenditure of ammunition to register their guns. As a result, many of the shells landed well behind the intended objective. Wire went uncut and machine gun emplacements went unscathed. Consequently, the Tsar's soldiers advanced against untouched fortifications.

Another problem that Stavka had came from a lack of tactical understanding. Still mired in the concepts of sweeping corps movements, Russian generals brought together large quantities of troops and material in preparation for an assault. Since the Germans controlled the skies, these concentrations were easily detected and von Hindenburg's staff could plan the movement of reinforcements or launch a pre-emptive artillery barrage which would destroy stores and demoralize the soldiers. Into this apparent void of tactical experience came General Alexei Brusilov, a man with little battle experience but an almost natural ability for adapting to the changing battlefield.

Brusilov was born into an aristocratic and military family in 1853. Groomed for the cavalry, where he spent most of his career, Brusilov never managed to garner a combat command. Instead, the majority of his duties were in staff assignments. As a captain and a major during the Russo-Turkish War, he was shielded by the administrative aspects of running a unit engaged at the front. In the Russo-Japanese War, he had remained behind as a strategic reserve against any European nation that might have taken advantage of Russia's preoccupation in the Far East.[31] Physically, he lacked the attributes of pomp and circumstances so often associated with Tsarist generals. The British General Alfred Knox, who was both an observer and an adviser to the Russian army from 1914 to 1917, described Brusilov laconically as having small, deep eyes and thin lips, intelligent, and self-reliant.[32] One of his many publicity photographs showed him sitting at his desk looking intensely at a spot just to the side of the camera which took the picture. His thin body gave the impression of a wiry, active man while his face looked as if it were cut from granite. The handle bar mustache that graced these finely chiseled lines appeared to be out of place, perhaps having been painted there by some passing sidewalk artist. Unlike his contemporaries, he wore very few medals, which gave the impression of the self-reliance General Knox had talked about. In 1916, he saw himself as the catalyst by which the Russian Empire could win the war.

Newly promoted to the position of commander of the Southwest Front, Brusilov proposed, in March 1916, that the Russian army should strike the Austro-Hungarians and Germans along the entire front from Riga to the Rumanian border. Initially, Stavka tabled the plan based on the objections of Brusilov's fellow front commanders. Generals Aleksei Evert and Aleksei Kuropatkin commanded the fronts immediately to Brusilov's right. Both had commanded in the Russo-Japanese War and suffered humiliation after the defeat. Given a second chance at command, neither wanted to commit to any plan which might lead to another failure.[33] Both generals cited the fact that repeated attacks on the Central Powers' defenses had been futile. In their estimation, the trenches were impregnable.[34] Brusilov was able to outmaneuver their protests through achieving the Tsar's support. Additional support came from the French who urgently asked for an immediate Russian offensive to relieve pressure at Verdun.

Originally scheduled for June 15, the offensive's main thrust would be directed at Lutsk and Kowel. Secondary attacks would occur opposite Brest-Litovsk, Lemberg, Stanislau, and Czernovitz.[35] Brusilov's plan not only abandoned the concept of attacking on a narrow front, it also envisioned having more than one objective in the offensive.

In May, French officials asked Stavka to move up the offensive's date. Things were going bad at Verdun. Stavka asked the area generals to advance the date to June 1. Brusilov agreed to the change but Kuropatkin asked for a delay until the fifteenth while Evert agreed to follow at a later unspecified date. Evert was still replacing losses from an abortive offensive near Lake Narotch between March 18 and 26 which had cost upwards to 100,000 casualties. Brusilov was still meeting his objective of striking along a broadened front but, initially, the front was only 250 miles long as opposed to the thousands that he had envisioned. Nevertheless, staggered start dates would still keep the enemies off balance and deter concentrated reinforcements to any part of the front. To ease Brusilov's misgivings about starting before the other front commanders, Stavka promised him reinforcements if they were needed.[36]

Brusilov's planning for the offensive was meticulous. To shield troop and material concentrations, he dispersed his units and supplies over the entire Galician Front.[37] This confused Central Powers' intelligence officers who had become accustomed to the Russians building large staging areas. They knew that a spring offensive was being planned but they were unable to pin down where the main thrust would occur. In addition to this dispersion, his staff used aerial photographs taken of Austro-Hungarian lines to build models far to the rear of the front. These fortifications were used to train the newly arrived enlistees and conscripts who had been raised in 1915. These new soldiers were not unlike the English Field Marshal Horatio Kitchener's Mob that had been raised to replace the depleted ranks of the 1914 British army. They came to the front with a strong sense of patriotism as well as a better education. But, by far, the greatest help to Brusilov's plans came from the Central Powers themselves.

In February 1916, von Falkenhayn began the Verdun campaign which effectively tied down German units on the Western Front and called for the movement of some troops and artillery off the Eastern Front. However, the key ingredient came when Field Marshal Conrad von Hotzendorf, noting that the inactivity on the Eastern Front was a good time for Austro-Hungarian troops to attack on the Italian Front, pulled fourteen battle-experienced divisions away from Galicia.[38] In effect, the Central Powers had placed themselves in a strategically precarious position should France or Italy ask the Russians to relieve the pressure on their fronts by attacking in the East. *OHL* would be unable to reinforce from the West as Austria-Hungary would be unable to move troops off the Italian Front without losing gains or jeopardizing themselves to successful counterattacks.

On June 4, German units were continuing their fight to recover Vaux Fortress in the Battle of Verdun. The effort to bleed France white had entered its fourth month. In the Ypres Salient, the Third Battle of Ypres was a day old. The Canadians had counterattacked to regain the approximately 300 yards lost the previous day. While on the Italian Front, Austro-Hungarian contingents continued to throw themselves against Italian strong points from Posina to Astico with little success. At dawn of that day, nearly 500 pieces of Russian artillery opened fire along a 200-mile front, raining shells on the Austro-Hungarian Fourth, Second, and Seventh Armies, and the composite *Südarmee*. Brusilov's offensive had begun.[39]

The barrage continued through the day and into the next. Daylight bombardment concentrated on the trenches and immediate rear areas to deter reinforcements and resupply while night firing shifted to no man's land to cut wire and to keep repair details out of the wire. At 12:30 in the afternoon of the second day the fire was lifted. Austro-Hungarian troops busily manned their positions, expecting an attack. It never came. Instead, Russian scouts and airplanes estimated the effectiveness of the artillery fire and called for more rounds to fall in among the wire. When the bombardment began again at 2:30, the Dual Empire's troops were massed in the forward trenches. The cannonade was devastating. Telephone communication with the rear areas and artillery support was severed. Consequently, Austro-Hungarian counterartillery measures went unguided and eventually ceased altogether.[40] On the morning of the sixth, the artillery fire finally stopped and the Russian army advanced. Within hours, the first three lines of trenches of the Austrian Seventh and Fourth Armies were captured. Gaps occurred in their lines in more than 20 places. Between the two faltering armies, the Austrian Second and the *Südarmee* held their positions, actually repulsing the attackers.

After 48 hours, a 50-mile gap existed in the Austro-Hungarian line around Lutsk. The Russian Eighth Army reported taking 40,000 prisoners in that time from the Dual Monarchy's Fourth Army.[41] By the end of the first week, an estimated 70,000 prisoners were taken along the entire front.⁴² Lutsk, the main objective of the offensive, fell during the second day of the advance. It was apparent that the Austrian Fourth Army was shattered and the Seventh was completely broken.⁴³ However, where there were Germans in the trenches, the line held. Both the Second and *Südarmee*, in the middle of the line, continued to hold; but, the utter rout of their neighboring armies began to leave their flanks exposed. They were obliged to retreat approximately 20 miles to maintain some semblance of contact with the remains of the Fourth and Seventh Armies.⁴⁴

On July 15, the Austro-Hungarian Fourth Army attempted a counteroffensive to slow the Russians' advance across the Lemberg-Brody railway. However, the massing of 20 divisions was inadequately concealed and the Russians launched a preemptive strike. Within a few days an additional 13,000 prisoners were taken and the way was open for a crossing of the Styr on July 21. Since the beginning of the offensive, the Russian Third and Eighth Armies had advanced as much as 75 miles.

So demoralizing and devastating was the offensive that Archduke Karl and Field Marshal Conrad von Hotzendorf were beginning to consider a separate peace with Russia.⁴⁵ Advocates of Hungarian independence also talked of a separate peace with Russia regardless of German or Austrian sentiments.⁴⁶ The Tsar's armies occupied the majority of Bukovina and were beginning to

A multinational force, Germans and Turks in Galicia

Author's Collection

assault the Carpathian passes when the Russian soldier's worst enemies, ineptitude and fear at the Stavka level, began to take a toll.

The first setback came from the north. General Evert, the front commander to the immediate north of Brusilov's front, postponed the planned diversionary attack toward Brest-Litovsk which would have kept German reinforcements away from the Galician gains. Further north, General Kuropatkin decided not to participate at all. He preferred to have his soldiers conduct ambushes and raids while maintaining a defensive posture before Riga and Dvinsk. With no threat coming from these areas, General Ludendorff, with the able assistance of General Hoffman, began moving troops south to reinforce the failing Austro-Hungarians. Assistance from the West in great numbers wasn't possible because of the renewed British attack on the Somme. Ludendorff could rely only on what assets he could spare from other areas on the Eastern Front. This amounted to five infantry and two cavalry divisions. To augment these meager forces, he was able to coerce Conrad von Hotzendorf into bringing seven of his divisions back from the Italian Front. As an added bonus, Enver Pasha agreed to send two Turkish divisions.[47] Belatedly, *OHL* promised an additional six divisions from the West when and if it was possible. With such massive amounts of manpower on its way, Hoffman could only hope for additional time. Moving reinforcements along the Galician Front was a problem. Only one railroad existed which could handle such monumental movements. By mid-July, Russian units were within 25 miles of this railroad as well as straddling it just above the Pruth River. This made movement along the line extremely difficult. A respite from the Russian onslaught for the Central Powers came in the very same way as rest from von Mackensen's advance had come for the Russians in September 1915.

The ardor that the highly trained Russian soldier of 1916 showed in the opening days of the offensive was extremely noteworthy. The capture of more than 100,000 prisoners and almost 75 miles of the enemy's territory was a just reward for the iron discipline exacted. However, Brusilov began to pay the price. Often *élan* in battle gives way to recklessness. Casualties to gunfire in the X Corps alone from July 7 to 14 had totaled 8,000.[48] Brusilov's reserves were running out all along the line. The requests to other front commanders for reinforcements met with postponements or flat refusals. Even at Stavka, despite prior promises to reinforce when necessary, help was only begrudgingly given. Grand Duke Alexeev acceded to sending the uncommitted Guard Army to Brusilov.[49] He was better off without them.

The Guard Army was the elite of the elite. Most of the soldiers were towering giants of over six feet and better equipped than other units. At 134,000 strong,[50] the army was split into two corps each of which was commanded by an arrogant, self-centered general who refused to obey any orders except their own. Working under such conditions, Brusilov was barely able to direct the army toward taking Kowel. Kowel was the center of the rail traffic for the entire Galician Front. If it fell, Ludendorff would not be able to speedily reinforce Lemberg or Stanislau.

The Guard advanced on Kowel through a marsh on July 25 against the remains of the Austro-Hungarian Fourth Army reinforced by two German divisions. Within a few days, the villages of Raemyesto, Shchurin, and Tristen fell. The Guard took 11,000 prisoners, 46 guns, and 65 machine guns.[51] For a time, it appeared that success would be repeated in this area also. But as the Guard advanced through the marsh—in places the water was chest deep—the Germans decided to expose them to the reality of total war.

German planes began strafing runs on the advancing troops. Antiaircraft fire could not protect them because of limited solid ground on which defensive machine guns or cannon could be placed. Those not killed directly but only wounded, drowned. From July 29 to 30, aircraft fire inflicted more than 30,000 deaths on the 2 and 3 Regiments.[52] The advance on Kowel was delayed while the two defiant corps commanders brought up reinforcements as did the Germans. Nine German battalions bolstered the Kowel fortifications along with additional artillery and machine guns. When the Guard began the advance again, they were met with devastating fire. By August 9, Guard losses amounted to 532 officers and 54,770 men.[53] Kowel had become a Russian Verdun, and the cream of the Russian army was cut down in droves to drown in stagnant water. All along the newly established front, Brusilov began to realize how alone he really was in this offensive. Reinforcements were not coming to his aid.

In the south, the Russian Ninth Army stood at the foot of the Carpathians but lacked sufficient reserves to push through the passes. To take pressure away from the Guard Army, Brusilov decided to commit what little reserves that were available to an attack on the Austro-Hungarian Second Army and the *Südarmee*. The objective this time would be the rail hub at Lemberg. The attack began on August 7 by both Eighth and Eleventh Armies. Initially, the Second and *Südarmee* began a slow retreat in the face of overwhelming forces. Brody and Halicz fell within a few days but the retreat remained organized and effective in keeping the Russians at a slow pace. The delaying action provided additional time to bring in reinforcements. *OHL* and Conrad von Hotzendorf had moved seven divisions into the Galician area from the west and Italian fronts.[54] Reinforced not only by troops from the west and Turkey, the *Südarmee* turned and stood on the Zlota Lipa approximately 30 miles from Lemberg. Faced with a well-organized frontal defense, Brusilov's offensive came to a stop. It was the last successful offensive of the Tsar's armies and the most costly.

REORGANIZATION OF CENTRAL POWERS, SEEDS OF RUSSIAN REVOLUTIONARY DISSENT

For Russia, the most glaring result of the Brusilov offensive was that the Tsar's soldiers had lost confidence in their leaders. General Knox reported an incident in which he was questioned by Russian soldiers as to why France and England had allowed Germany to move whole divisions from the West. His reply was that the Germans had replaced those divisions with fresher, newly raised divisions. Knox showed them that there had been 115 and a half German divisions on January 1 and that by October 22, this number had increased to

127 divisions.⁵⁵ His answer had been calculated to show that Russia was not "in it alone" as so many Russian soldiers thought. England and France were doing their best to tie down as many divisions in the West as they could, but the Germans were just as numerous as they had always been. To the Russian soldier, this response undoubtedly convinced him that he was fighting a war that would go on forever. Brusilov's offensive, although successful, cost more than 500,000 casualties. For replacements, the Tsar's agents began impressing older, married men. Torn away from their rural villages and families, these inductees did not display the same zeal and patriotic sentiment of those recruits who had led the armies deep into Galicia and Bukovina. Nor could these people be trained to the degree of effectiveness that the others had been. Because of the fragileness of the newly established line, they had to be launched into the trenches with only minimal training. Statistics began to show the dropping of effectiveness. From October 1 through December 31, there was an increase of 136,089 taken as prisoners by German and Austro-Hungarian soldiers while casualties began a marked decrease of almost 200,000 during the same period.⁵⁶ To add salt to this wound, the entry of Rumania on the *Entente*'s side stripped away two and one half corps from the front and drained much needed materials.

For Germany's *OHL*, the Brusilov Offensive had finally given them the opportunity to establish hegemony on the Eastern Front. Since the winter campaigns of 1914–1915, *OHL* had been attempting to gain supremacy on the Front. They had pointed to their continuing reinforcement of the Austro-Hungarian forces as well as the failure of Conrad von Hotzendorf's many offensives as reasons for consolidating command. Emperor Franz Joseph, ably counseled by Conrad, had been able to fend off the many demands for a unified command, but the success of the Russians in June and July had made the Germans more demanding. In the early days of the enemy's advance into Galicia, *OHL* had suggested an Eastern Front commander in the way of Field Marshal von Mackensen, but the crisis was too deep to warrant immediate attention. It was not until the fall of Brody, that Franz Joseph was willing to concede to *OHL*'s demands. Through negotiations, the aging emperor agreed to a German supreme commander for the Eastern Front in the personage of Field Marshal von Hindenburg. All Austro-Hungarian Armies with the exception of the Third Army and the *Südarmee* operating on the extreme southern flank would fall under German direction. Although nominally independent in their actions, these two armies were really under German control. The Third Army was commanded by Archduke Karl, heir apparent to the empire's throne, while his chief of staff was the German General Hans von Seeckt. Direction of the army's strategy was in von Seeckt's hands. The *Südarmee* remained under the direction of a German commander. Politically, economically, and militarily, the Dual Empire was no longer its own master.

CHAPTER THREE
Rumania

The year of 1916 is known for the bloodshed of Verdun and the Somme. These were fantastic battles which claimed, in come cases, 60,000 casualties in the space of a few minutes. In both battles, the outcome of the carnage was indecisive except to those who gave up their lives. However, for scholars of the Western Front, the engagements always are cited as decisive in that they signaled a realization by the German public and soldier that the war could no longer be won by military force. On the Eastern Front, the events of 1916 had the opposite effect. Successes on the Baltic, in Galicia, and against Rumania were making the Reich's dream of an economic power in *Mittel Europa* a reality.

Just the year before, German soldiers, commanded by Field Marshal August von Mackensen, had scored weighty victories in Russia and Serbia. In the case of Serbia, Austro-Hungarian armies had suffered two humiliating defeats in attempts to put an end to that nation's belligerency. German know-how, material, and troops had bolstered their ally's sagging forces and, in a whirlwind of artillery shells and machine gun bullets, Serbia had fallen like a ripe plum into the Central Powers' laps. Although not as decisive as the Serbian victory, the Germans' triumphs against Russia had been of strategic value. The *Reich*'s armies had advanced with unequaled speed across the Polish plain driving the Russian steamroller in reverse. The advance had stopped only as a result of overextended supply lines and exhaustion.

Despite these massive victories, the Eastern Front was beginning to resemble the Western theatre. German and Austro-Hungarian forces were entrenching themselves on a line extending from Riga to the Rumanian Carpathians. What was needed by both the *Entente* and the Central Powers to preclude a second front of attrition was a way to turn the flank. The Baltic prevented a

northern approach. The only feasible direction was from the south. And, on the south lay the still neutral kingdom of Rumania.

ALLIANCES, INFLUENCES, NEUTRALITY

Rumania's borders had placed it in an enviable position. Made up of the two principalities of Wallachia and Moldavia, Rumania appeared to be opened scissors waiting to cut into the underbelly of the Austro-Hungarian Empire. In the angles of its scissors shape stood the high Transylvanian Alps with its passes between the Hungarian plains and Wallachian lowlands firmly in Rumanian hands. To the south was the mighty Danube River which, at one point, had a breadth of more than 1,300 yards. And to the east, Moldavia was bordered by the Pruth River. Only in the east had Rumania seen a need to add a string of fortifications to her natural barriers.[1] Even the newest piece of Rumania, the Dobrudja province, which bordered the Black Sea, was not without its difficulties. It was a veritable bog watered by the Danube's delta.[2] If either the *Entente* or the Central Powers forced war on Rumania, the country stood ready to defend itself behind these formidable natural borders. But in some instances defenses are also used to keep the world from knowing what is going on inside.

Rumania's king in August 1914 was the much-respected Carol I who had ascended to the hereditary throne in 1881. He was head of the house of Hohenzollern-Sigmaringen, a branch of the ruling house of Prussia and Germany. Consequently, it was no secret in which direction the king would steer Rumanian troops if the parliament had allowed him. Citing a secret defensive treaty he had signed with Germany and Austria-Hungary in 1883 as part of Bismarck's grand scheme of alliances, Carol had attempted to conjure his ministers and parliament into siding with the Central Powers.[3] But the business of government had changed since the late nineteenth century; kings no longer ruled absolutely. The liberal government led by Iosef Bratianu saw a different objective for Rumania.

Periodically, historians attribute World War I's cause to an awakening sense of imagined or real nationality. For Bratianu and his ministers and the Rumanian people, a firm belief in ethnic integrity fueled their ambitions. The ministers desired to acquire the area that was traditionally seen as the homeland of the Rumanians, Transylvania, which was a large part of the Austro-Hungarian Empire.[4] The oppression of Rumanians by the Hungarians, including attempts to stifle the language and customs, had fed Rumanians' hostilities toward the Dual Empire over the decades. Carol had attempted to shift the people's interest from Transylvania to the Russian province of Bessarabia, also inhabited predominantly by Rumanians, but his efforts had fallen on deaf ears.[5]

Alienated from both his ministers and his subjects, Carol threatened abdication if Rumania entered the war on the *Entente*'s side. This was a crisis which the ministers were not inclined to face. Despite their adherence to new, twentieth-century governmental philosophies, they were not willing to part with their monarch. To avert a disturbance, the king and the ministers agreed on a path of neutrality.

Neutrality served both the king's and the ministers' aims. In August 1914, very few people were intuitive enough to know that the war would last for four years. Carol undoubtedly saw an easy victory for the German and Austro-Hungarian Empires. The Rumanian ministers probably reckoned that French and Russian forces would handily defeat the Germanic empires. Italy's refusal to join the war, thus limiting pressure on France as well as Austria-Hungary, furthered each side's belief that the war would remain small and short. For a few months, the deadlock between the interests remained when, not unexpectedly, Carol died and his nephew, Ferdinand, ascended the throne.

Ferdinand was of a different cut than Carol. No longer considering himself descended from Germanic stock, he adamantly described himself as being a Rumanian. As a nationalist, his views were similar to those of Bratianu and his ministers. All agreed that Transylvania had to be united with Wallachia and Moldavia.[6] Yet Ferdinand and Bratianu were very cautious. Although England and France had stopped Germany on the Marne, her troops had soundly defeated the Russians at Tannenberg and the Masurian Lakes. Even Austria-Hungary, suffering initial defeats, had managed to rally and hold the Russians at bay in Galicia and Bukovina. Rumania's best course was to remain neutral until both sides had worn themselves down and then enter on the side of the stronger. For a while, Rumania would play the saucy coquette allowing both the Entente and the Central Powers to lavish favors on it in hopes of eventually bringing about an alliance.

Germany tried to bring Rumania into the war based on the existing secret treaty that Carol had cited. When Bratianu ignored those treaty obligations, the Reich, as in the case with Italy, pressed Austria-Hungary to meet some terms which would either continue Rumanian neutrality or bring it into the Central Powers' camps.[7] Those terms included offering Russian Bessarabia and political reforms in Transylvania. Hungarian interests, caught up in their own nationalistic fervor, were willing to offer Bessarabia but categorically refused to offer any concession to political rights. The *Entente* pursued a different direction.

In the autumn of 1914, Russia made its first proposal not on behalf of the *Entente* but rather for its own security. In return for continued neutrality, Russia agreed to the Rumanian right to annex Austro-Hungarian territory inhabited by the Rumanian majority.[8] Bratianu was being offered territory without having to fight for it. This was in sharp contrast to the Central Powers' offer which would have required Rumania to enter the war and maintain hostilities with Russia to gain Bessarabia. Considering Rumania's approach of waiting until all the belligerents spent themselves before choosing sides as they had done in the Second Balkan War, the Russian offer to maintain neutrality was the better of the two bids. Bratianu therefore signed a neutrality treaty with Russia on October 2. Contrary to what some authors have espoused, the agreement to continued neutrality cannot be wholly interpreted as an indication of where Rumanian interests would lead.[9] It is true that Rumania's designs were on the dismemberment of Austria-Hungary, but these intentions did not need to come about through armed conflict. Diplomacy was a far less expensive tool. It appears that Rumania intended to bring about its territorial aspirations without picking a side.

Throughout 1915, Rumania maintained its coquettish role. Preoccupied with successes in Poland, Germany allowed Austria-Hungary to court Rumania. Hungarian interests laid little on the front door's offering table. Consequently, Bratianu conducted serious negotiations with the *Entente* through the back door. In May, they even reached a tentative accord by which Rumania would receive all of Transylvania, Bukovina, and the Banat and the Hungarian counties along the Pruth River[10] in return for their entry into the war. But when the Central Powers quickly and decisively executed Serbia, Bratianu refused to entertain any thoughts of intervention. Instead, Rumania chose to show benevolence to both sides by opening her vast stores of grain and oil and refusing passage across Rumania of Russian soldiers for the beleaguered Serbians.[11]

Feinted favors were no longer stylish in the world of 1916. The Germans had suffered at Verdun while the Austro-Hungarians fled before the Russians' offensives in Galicia. The *Entente* had failed at Galipoli and suffered losses on the Western Front equal to those of the Germans. For both sides, Rumania had to either make a decision or be forced into one.

The Rumanians were equally desirous to end their teasing ways. Russian advances into Bukovina had been more successful than were hoped. The Tsar's soldiers stood on the summits of the Carpathians looking down on the Hungarian plain. Any further advances would bring Russians into Transylvania. The question would then arise about whether or not the Russians would vacate Transylvania after they had shed their blood to occupy it. Rumania could lose the coveted price they were exacting for continued neutrality to Russian conquest. Rumors were already circulating that Budapest was considering a peace initiative of their own.[12] The time was ripe to leap into the fray.

On July 4, Bratianu listed the final cost for Rumanian entry on the *Entente*'s side: continuation of the English and French offensives on the Western Front,[13] stabilization of the Russian advance into Bukovina and Galicia, security against attack from Bulgaria through an assault out of Salonika and 200,000 Russian troops in the Dobrudja,[14] a supply of 300 tons of munitions a day,[15] and, finally, agreement to the territorial demands of 1915. The negotiators agreed to buy but, in a last fit of coquettishness, Rumania upped the price ever so slightly by asking that all territories that were not conquered by the end of the war would be subject to "general conditions."[16] In this way, Rumania intended to ensure that all territorial demands were turned over to them regardless of whose army stood in occupation of an area.

This last cost addition was unacceptable to the *Entente* who saw Russia and Serbia losing any hard-won territory, but the disagreement to the price was not conveyed to Bratianu. Instead, the Russians and French secretly agreed that "the annexations promised to Rumania will be effective only as the general situation permits."[17] As great powers in the war, they "reserved for themselves the big questions."[18] In effect, they agreed that the peace table was the place to discuss appropriations. Rumania would have to make good its entry into hostilities with blood if it expected territorial gains. On August 9, 1916, the *Entente* sent word to Bratianu that his terms were acceptable and demanded Rumania declare war on Austria-Hungary. In return for agreeing to the price of intervention, the *Entente* expected Rumania to open hostilities within eight days after the beginning of the Salonika offensive,[19] cease all commerce and diplomatic relations with Germany, Turkey, and Bulgaria, and not to enter into a separate peace with any of the Central Powers.

There must have been a fair amount of elation among the Rumanian ministers. Diplomatically, it appeared that they had carried off a fantastic work of salesmanship. They had gained recognition of territorial demands as well as support and protection against Bulgarian invasion. Now the trick was to bring about a victory through blood and steel. This was something that the Rumanians were not as well prepared for as were the Central Powers.

The Central Powers were not entirely taken in with Rumania's false mantle of neutrality. Rumors about the negotiations between the *Entente* and Rumania abounded in Rome.[20] Germany reacted to these innuendoes by pressuring Austria-Hungary into a final effort to wean Bratianu away from the *Entente*. The offer of all of Bukovina, undoubtedly coming too late and too small in proportion compared to what the *Entente* offered, was never accepted.[21] By the end of August, there could be no doubt that Bratianu was about to bring Rumania into the war. In a *faux pas*, Bratianu attempted to negotiate a neutrality treaty with Bulgaria.[22] Germanic ideals took over. The Central Powers responded to the proposal on behalf of Bulgaria by saying that any attack by Rumania on Austria-Hungary would put Rumania at war with all of the Central Powers.[23]

SALONIKA, GALICIA, AND THE RUMANIAN ARMY

There were three items which King Ferdinand and Bratianu placed their trust in for a quick and easy victory over Austria-Hungary. The first was a sustained *Entente* offensive out of Salonika which would pin down Bulgaria's armies, thus decreasing the likelihood of an attack across the Danube. Second, the cessation of the Russian advance into Bukovina would mean a free hand in Transylvania. And the third item was a reliance on the massive Rumanian army.

On the surface, the *Entente*'s presence at Salonika seemed to have come into existence as a result of the devastating attack on Serbia. The *Entente* had envisioned a force which would conduct a quick campaign into the Balkan peninsula to save Serbia or at least relieve the pressure of the Central Powers' invasion. However, before a sufficient force could be formed to effect one of these ends, Serbia had been occupied and its army decimated. With their objective no longer attainable, troop concentrations should have ceased or at the very least been repositioned at Gallipoli. No such order was given and soldiers continued to land at Salonika.

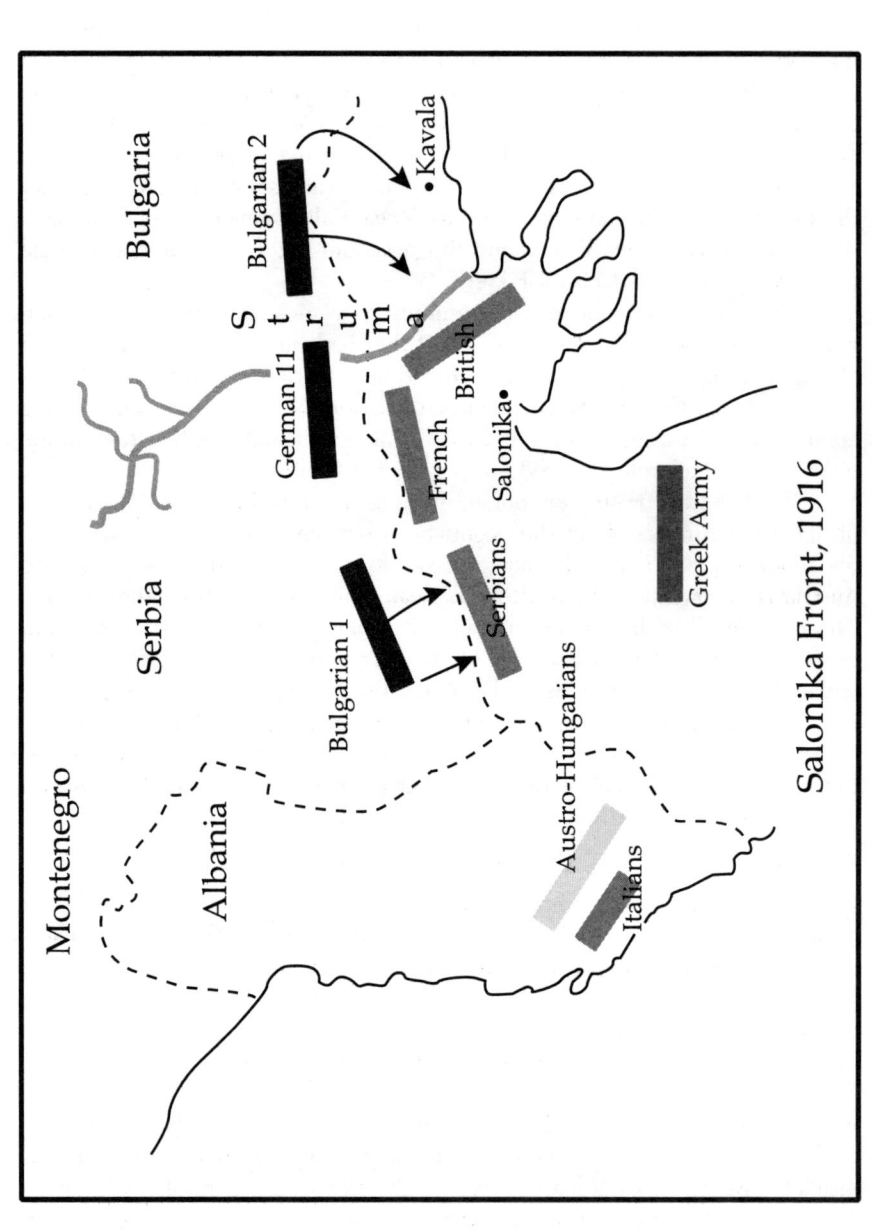

Commanded by the French General Maurice Sarrail,[24] the *Entente*'s forces at Salonika, termed as the Army of the Orient, had contingents from almost every nationality participating in the war. Besides French and British troops, there were Russians, Italians, and Serbians.[25] Sarrail's command consisted of nationalities which had political aims in the Balkans and Asia Minor. The Italians had designs for the Albanian coast. Russian presence was probably an insurance toward precluding anyone else from conquering the Bospherous. With such diverse strategic aims and the lack of a Serbian nation to rescue, it's little wonder that Sarrail's army had no clear-cut mission.

Sarrail, nevertheless, knew who the enemy might be. Based on the location of Salonika, and the fact that the only railroad servicing the port ran directly north into Bulgarian-held territory, the most likely objective was to open a front against Bulgaria. But Sarrail found it difficult to identify the enemy after the Greek government adopted a pro-German neutrality stance instead of declaring war against the Central Powers.[26] This placed his army amid a potentially hostile Greek army. Considering his strength, Sarrail had advanced only 60 miles from Salonika and then entrenched his territory choosing to adopt a wait and see attitude during which he could consolidate forces and supplies for a campaign in either direction.

By the summer of 1916, the Army of the Orient's strength and supplies were ready for an offensive. The negotiations between the *Entente*'s and Rumania's diplomats finally had fixed the army's objective as a front with Bulgaria. The Bulgarians felt equally confident about mounting a limited offensive against Salonika and did so two days before the *Entente*'s attack was to begin.[27] Ostentatiously, the offensive set out to flatten a salient extending from Florina on the west to the Vara River valley on the east.

Reinforced by two German divisions, Bulgarian units began the offensive on August 18 on the western fringes of the Kanli Valley against their old adversary, the Serbs.[28] Within hours, the station at Florina was captured. French and Serbian units at the center of the salient were thrown back. Further east, the duplicity of the Greek army which Sarrail had feared became evident when the Greeks surrendered the forts around Kavalla to the Bulgarians without resistance.[29] Flushed with such successes, the Bulgarians saw themselves in Salonika within a week. However, the momentum finally slowed on August 26 when Serbians, supported by the French, consolidated a defensive position on a curved western flank. Sarrail, smarting from a strategical and tactical shift in his line, wasn't able to pursue a counterattack until late September.

The Central Powers publicized the Bulgarians' achievements in the hopes of precluding Rumania's entry into the war. Whether the action at Salonika could have been termed as either a defense or offense, the Rumanians saw it as an occurrence which tied down Bulgarian forces. In fact, Bulgaria's army had attained their goal of flattening the front and were positioned on all the most advantageous points. Their presence could be reduced and shifted to the Danube with little degradation of the Salonika line. This deceptive appearance of events was also the case in Bukovina and Galicia.

Considering the vigor and resolve that the Tsarist army showed in its summer offensive of 1916, few people would have thought that it was the empire's swan song. After a short artillery barrage, the Russians had stormed forward into the Austro-Hungarian trenches. Within two weeks the Austro-Hungarian Fourth and Seventh Armies had collapsed, yielding over 100,000 prisoners and countless dead. Throughout the *Entente* nations, there was once again talk of the Russian "steamroller" carving a path of destruction to Berlin and Vienna within a few weeks. However, the nemeses of the Russian army, corruption and jealousy, took their toll. Without supporting attacks on other Russian fronts, OHL and Ludendorff were able to reposition surplus battalions from the Riga line and the Western Front in time to plug the gaps left by the defeated Dual Monarchy's armies.[30] By late July, the front was beginning to stabilize.

The German *OHL* was the real winner of the Russian summer offensive. Because of the disaster suffered by the Austro-Hungarian army, *OHL* was able to establish primacy in the theater. As payment for the rescue of the Front, *OHL* exacted Franz Joseph's approval of a commander in chief for the entire front. With the exception of one army which remained under the command of Archduke Karl, all units, regardless of nationality, henceforth took orders from Field Marshal Paul von Hindenburg.[31] This consolidation of actions cut much of the bureaucracy associated on the front. The Central Powers' counteroffensive became a coordinated action rather than one of cooperation and political maneuvering. As a result, the Central Powers were able to straighten the Galician line and secure key points which put Russian forces on the defensive rather than allow further advances.

As with Salonika, the *Entente* armies were not in a position to seriously support a new belligerent on their side. Rumania would have to rely on her own military prowess which, in 1916, was considered substantial by everyone except the *Entente*'s armies' commanders. King Ferdinand and his prime minister, Bratianu, placed a lot of faith in their army. The active force amounted to 250,000 infantry and 20,000 cavalry.[32] With reserves thrown in, the army could muster a strength of approximately 600,000.[33] As impressive as this manpower was, the Rumanian army was deficient in many areas. For artillery, they had 1,300 cannons of which not more than half were modern pieces from Krupp and Creusot.[34] The soldier carried a Mannlicher rifle made in Austria but there were not enough to equip all of the soldiers. However, the most glaring lack was in battle field experience.

PREPARATION, ATTACK, AND COUNTEROFFENSIVE

The German *OHL* was not as uninformed as the Rumanians or the *Entente* expected. By the end of July, von Falkenhayn was already aware of Bratianu's intentions to enter the war. Not only were there many rumors among Italian diplomats but statements made by Rumania's king and queen were enough to form a conclusion.[35] *OHL* was also aware of an arms shipment moving across Russia with a destination of Rumania.[36] A prudent course of action would have been to invade Rumania before their armies attacked, but the Russian advances in Bukovina and Galicia precluded new challenges. The next best thing to a pre-emptive strike was to plan for an eventuality of invasion and the resulting counteroffensive.[37]

From all indications, *OHL* suspected that Rumanian soldiers would invade Transylvania as soon as the harvest was complete.[38] The plan would be to allow the Rumanians to advance into Transylvania offering as stiff a defense as possible. The Rumanians' lack of transport as well as the mountainous terrain would present formidable delays of their own. Once the troops were committed to the northerly invasion, the Bulgarians would effect an invasion of Rumania from the south. With a relatively weak force of seven divisions, the Bulgarians would advance into the Dobrudja as well as cross the Danube at Nikopoli for an attack on Bucharest. Such a threat in the south would cause Rumania's generals to thin their invasion forces, transferring soldiers to stop the southerly advance. The time bought by a stiff resistance[39] along predisposed lines would enable the Central Powers to amass sufficient divisions for a counteroffensive against the weakened Rumanian line in Transylvania.

Using the mountains as natural parameters for an envelopment, the Central Powers' battalions would not only confront the invasion force but also sweep around its flanks to seal the mountain passes in its rear. Snows at passes in the higher elevations would limit any reinforcements from Rumania. The invasion would be smashed against the mountains by a frontal assault while the forces holding the passes would deter any escape. With the main part of the army gone, a counterinvasion of Rumania's northern province would follow. Eventually, this force would link up with the Bulgarians and administer the final coup de grace.

To lead the southern army, *OHL* selected Field Marshal von Mackensen. His army consisted of one German infantry division, Bulgarian forces, two Turkish infantry divisions, and an Austro-Hungarian Danube flotilla and bridging equipment.[40] Massing the divisions as well as the bridging equipment began in mid-August. The *Entente*'s intelligence gatherers soon caught wind of the von Mackensen placement and the troop concentrations. The event was translated into a possible attack against Salonika. The interpretation was probably reinforced when the Bulgarians launched their limited objective offensive on August 18. For the Central Powers, everything was in readiness for the Rumanians' entrance. The Rumanians needed only to step out.

Rumanian preparations for war began around August 16 with mobilization and the deployment of a Russian contingent consisting of two infantry divisions and one cavalry division in the Dobrudja. War was declared against Austria-Hungary on August 27 and the invasion of Transylvania began only a few hours later.

In a three-pronged drive, Rumanian forces quickly took all the mountain passes' exits. The First Army, approximately 135,000 strong, moved north out of Wallachia through the Vulcan and Red Tower Passes. The Second Army, 126,000 soldiers, advanced in a semicircular shape through the Torzburg, Buzer, and

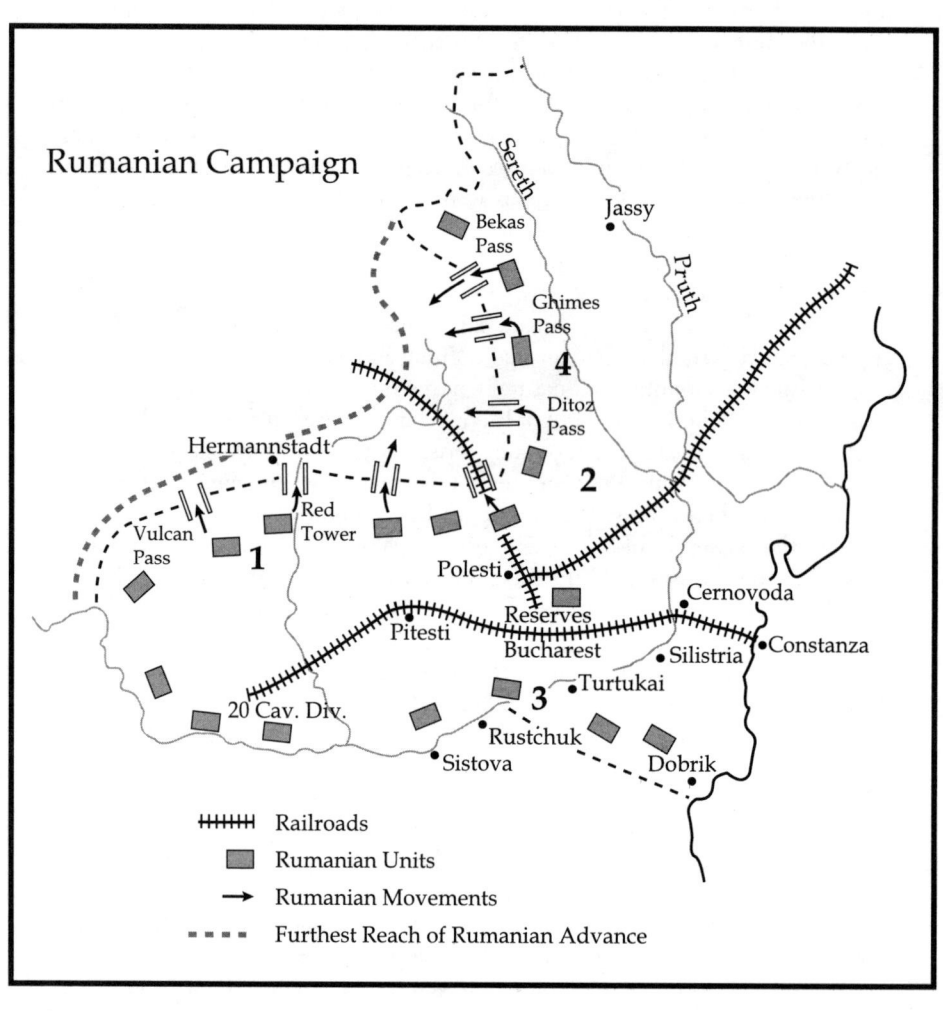

Ditoz Passes. The third prong, the Fourth Army consisting of 107,000, proceeded out of Moldavia through the Bekas and Ghimes Passes.[41] As planned, the Austro-Hungarian defenders withdrew in an organized rear guard action to prepared fortifications at the Maros and Little Kodel Rivers. By mid-September, all three armies began consolidating their positions along a front stretching from Dorna Watra in the north to Fagaros and Hermannstadt in the center to a few miles beyond the Vulcan pass in the south. A full one third of the coveted Transylvania was occupied. The Rumanian army's sweet taste of success in the north, however, slowly turned bitter as the Central Powers initiated their plan of countercampaign.

Field Marshal von Mackensen's Bulgarian-German force entered Rumania by way of the Dobrudja on September 1. Realistically, with only 70,000 troops at his disposal, von Mackensen could effect only limited objectives. Undoubtedly, his primary concern was to cause the thinning of the Rumanian armies in Transylvania. Even as important as that objective may seem, von Mackensen must have also had two other objectives. The first goal was the capture of the port of Constanza. This would bring about the completion of his second objective of drawing Russian troops off the Galician line in an effort to stifle any outflanking attack through Bessarabia.

Constanza had two very important characteristics for any invading army to consider as reasons for its capture. On the one hand, the bridge that ran from Constanza to Cernavoda was the only span which crossed the Danube between there and Belgrade.[42] But the reason on the other hand was much more important. Constanza was Rumania's only port on the Black Sea and the terminus of the only railroad leading from the sea to Bucharest.[43] Because of the lack of common railroads between Rumania and Russia, Constanza represented the best place to land supplies and Russian soldiers which had been transshipped across Russia. Take Constanza, and *Entente* support would be considerably reduced.

To guard their southern borders, the Rumanians had placed their Third Army along the Danube. Relying on their agreement with the *Entente* to provide 200,000 Russian soldiers to defend the Danube frontier, Rumania's generals had placed only small contingents in the Dobrudja. Defense rested primarily with the Russians. When the Tsar sent only 50,000 combatants, the Rumanians began thinning the Danube line areas. Approximately 142,000 strong,[44] the Third Army should have been able to hold their own against the Central Powers' invading army. However, very few of the soldiers had seen the lightning speed of German and Bulgarian artillery.

Von Mackensen's Bulgarian Third Army moved against Turtukai and Silistria to secure the flank in the opening days of the invasion. Turtukai was a fortified city with a substantial complement of defenders. Contemporary writers stated that most of the Rumanian contingent were raw recruits who were undergoing training. For them the trademark cannonade of advancing German armies must have been twice as horrifying than it would have been for seasoned professionals. One hour in duration, the bombardment preceded a frontal attack on the town. Fighting lasted a little longer than 24 hours. In comparison to actions on the Western Front, the assault could have been termed as limited in scope but, for such a short burst of activity, the Rumanian losses were catastrophic. The Central Powers captured 28,450 soldiers and 151 guns.[45] So devastating were the losses that General Alexandru Averescu, the commander of the Third Army, ordered the evacuation of Silistria, the Bulgarians' next objective, instead of suffering more losses.

Von Mackensen's units continued their advance up the Dobrudja, meeting only stiff resistance in the center of their line. The Rumanians, reinforced by the Russian contingent,[46] held a defensive position running from Tuzla to Tosova for five days.[47] To dislodge the Russo-Rumanian force, the Bulgarians used a classic ruse; they appeared to be retreating. The Russo-Rumanians attempted a pursuit but they were quickly outflanked and encircled. Luckily, the encircling Bulgarians were not strong enough to hold the enemy and most escaped.

After advancing more than 60 miles and fighting for five days, von Mackensen's soldiers found it necessary to consolidate before forging ahead. Bringing up supplies proved to be the most trying of tasks. Severe rains had turned the roads into a morass.[48] Food supplies, being brought in along the Danube by boat, had been sunk at Rustchuk.[49] To add to these difficulties, the Rumanians attempted a counterinvasion of Bulgaria.

Until adequate forces could be repositioned from Transylvania, General Averescu devised a plan to take pressure off the retreating Russo-Rumanian armies. Consolidating nearly 177 infantry battalions[50] supported by cavalry and artillery, the Rumanians attempted to cross the Danube at Turtukai and Rustchuk. From the beginning, the attack seemed doomed. High winds made it impossible to maintain a pontoon bridge across the Danube, forcing the troops to cross by boat. The high winds had also prohibited the Rumanians from laying protective measures upstream for the crossing. When the winds died down, the prepositioned Austro-Hungarian flotilla was able to attack the troop carriers repeatedly. Eventually, some soldiers were landed on the Bulgarian side of the river but they were unable to consolidate a beach head because of defending artillery and machine gun fire. At the end of one day, the Rumanians were back on their side of the river in defensive positions. Those who had managed to cross the wide Danube were rounded up and taken prisoner.

Von Mackensen's force was ready for a continued offensive by the middle of October. Bolstered by two Turkish infantry divisions,[51] the advance began on October 19 and succeeded, four days later, in capturing Constanza. Cernavoda fell two days later despite the destruction of the bridge linking the two cities. As before, most of the Russo-Rumanian force had been encircled but the Bulgarians were too weak to keep them contained. Many Rumanians escaped across the destroyed bridge while the Russians retreated toward Russia, finally managing a consolidation 12 miles outside of Constanza. By this time, Rumanian soldiers were beginning to arrive from Transylvania. The ultimate objective had been met; the Rumanians had thinned their units in the north. The Central Powers' second offensive could now begin.

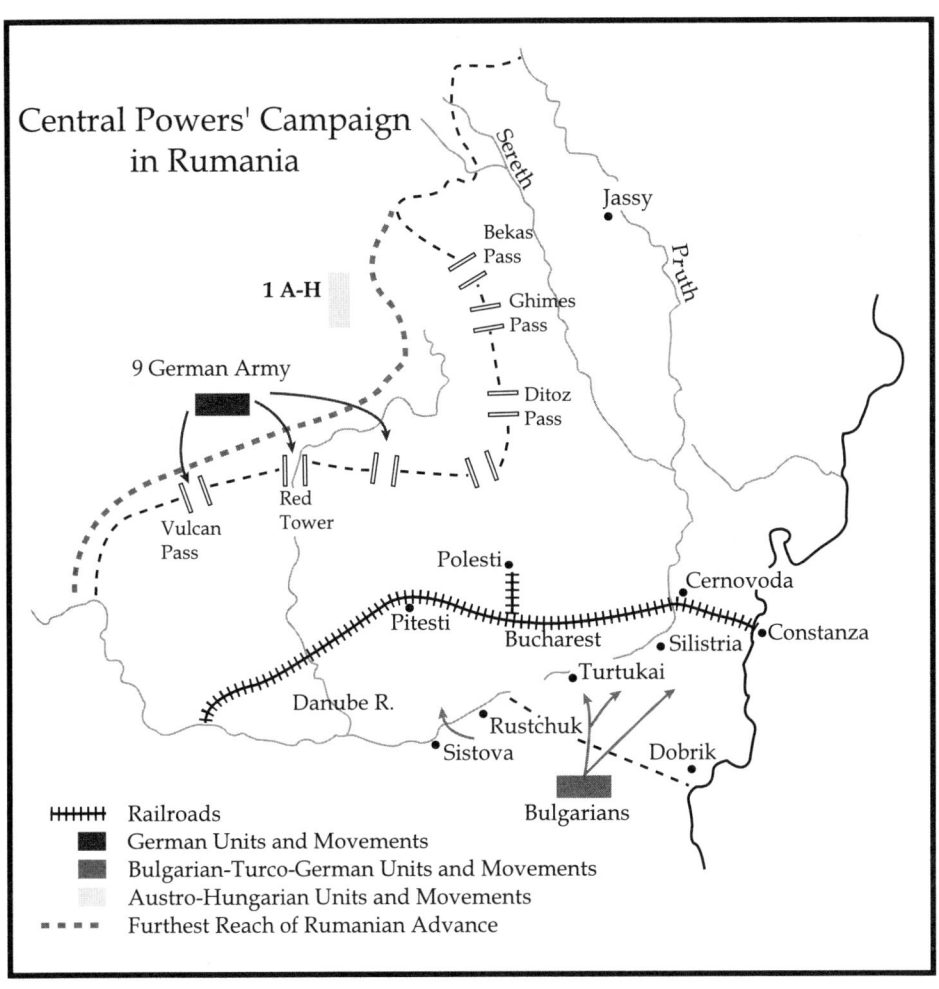

Rumania's entrance in the war had come as no surprise to the *Kaiser* or Bethmann-Hollweg. What had stunned them was the timing. Contrary to von Falkenhayn's estimation, the Rumanians had not waited for the harvest. This miscalculation, coupled with the debacles at Verdun and on the Somme, destroyed the Kaiser's confidence in von Falkenhayn and led to his dismissal and replacement with the dynamic personalities of von Hindenburg and his alter ego Ludendorff. But von Falkenhayn was not to be cast off as useless scrap. In September 1916, Erich von Falkenhayn found himself in command of the Ninth Army[52] and charged with the liberation of Transylvania and the counterinvasion of Rumania from the north. Because of his intimate knowledge of the Pless meeting in July 1916, there could have been no better choice.

By the beginning of October, Rumanian forces had stopped their advance into Transylvania. The movement of battalions to the Dobrudja to stop von Mackensen's invasion had sufficiently depleted the armies, requiring them to go on the defensive except in the extreme north. Remaining in Transylvania were ten infantry divisions, and one division and four brigades of cavalry.[53] The Central Powers' Transylvania offensive began on October 3 in a three-pronged drive. The southern prong, consisting of the crack Alpenkorps, made a dash to the Red Tower Pass attempting to cut off any retreat through it. In the north, the VI Corps made a frontal assault all along the Rumanian Fourth Army line. Despite Russian reinforcement, the Rumanians were forced back through the Bekas and Grimes passes. The center prong, however, was the main attack.

The battle at Hermannstadt opened on September 26 with the customary artillery barrage against the Rumanian I Corps. Although meager at 54 batteries when compared to von Mackensen's cannonade at Tarnow and on the San in 1915, the Rumanians' reply from 16 field batteries and two 120mm batteries was easily missed among the thunder of the German batteries. The cannonade continued through the day against the I Corps' frontal positions while the *Alpenkorps* moved around the western flank of the corps in an attempt to take and seal Red Tower Pass. Despite the diversion in front, the Rumanian commanders dedicated the *Korps'* movement and sent reinforcements to assist those units holding the pass. This was a sound tactical move but had ramifications in the next day's fighting. By the end of the first day, the Germans had successfully probed the Rumanian positions in front but were unable to take the pass behind. When the second day of battle dawned, the Rumanians still held the pass although command of the road leading to it belonged to the *Alpenkorps*. Von Falkenhayn sent in the infantry.

The 187, 51, and 76R regiments were launched in a frontal attack on the I Corps. As the fighting wore on, it became clear to the Rumanians that additional reinforcements were needed. Throwing three divisions and two cavalry brigades into the flank of von Falkenhayn's force seemed to lower pressure on the I Corps' front. However, the 1 and 3 Austro-Hungarian Regiments managed to give the illusion of pulling back when actually they were engaging the force effectively to deter their rescue of the I Corps.

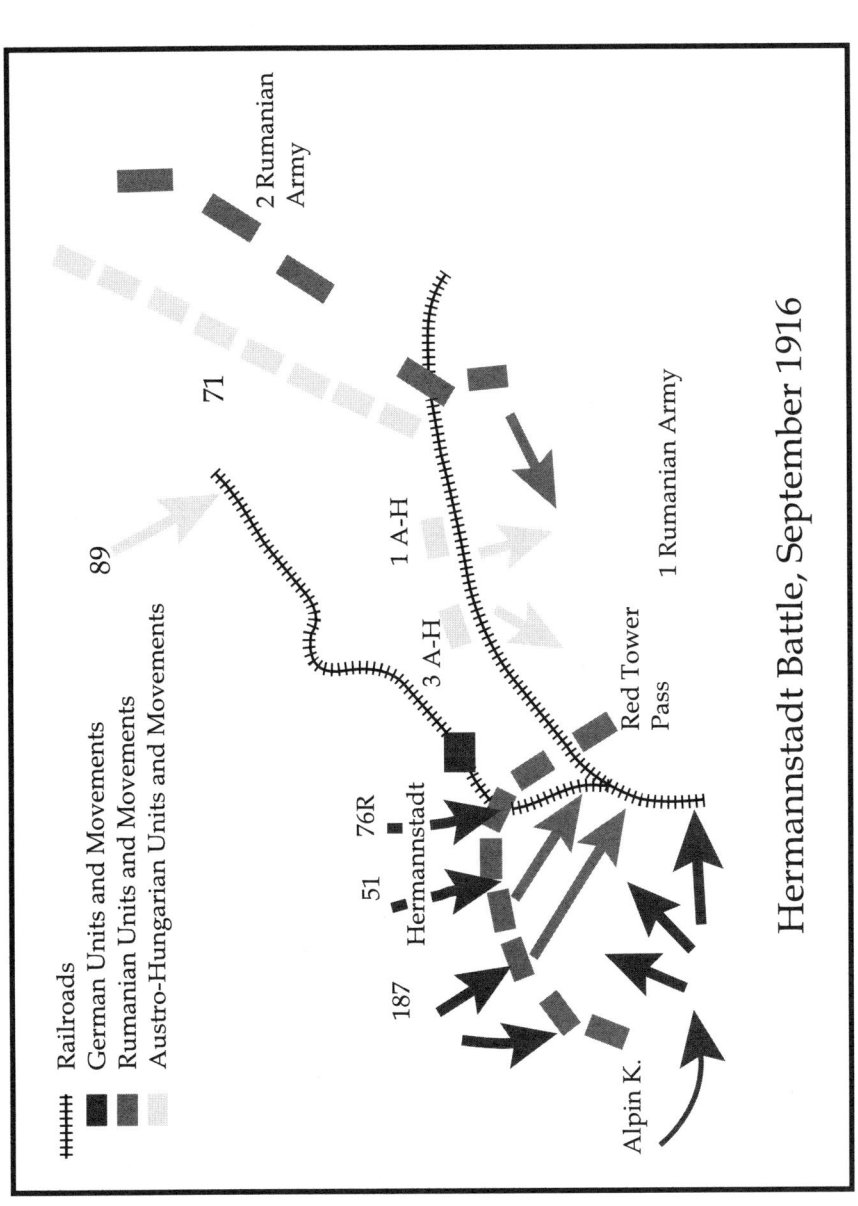

By nightfall on the third day, September 29, the I Corps was crowded around the northern exit of the Red Tower Pass. Artillery batteries were moved up, and firing began with devastating effect. Units began to break up into small groups which felt their way along mountain paths back into Rumania. Strategically, the route of the I Corps left the Second Army's flank dangerously exposed, causing it to begin a withdrawal to the passes and the perceived safety of Rumania.

Jubilant, von Falkenhayn reported to the Kaiser that after a three-day battle, the Ninth Army had "utterly crushed"[54] the Rumanians. An assessment of the battle's result does not bear out the victory report. After three days of heavy fighting the Central Powers soldiers accounted for only 3,000 Rumanian prisoners and 13 captured guns. Von Falkenhayn's report in fact stated that the Rumanian First Army was no longer a cohesive group and that the 2 Army was retreating despite having borne little punishment. By October 10, the Rumanian armies were back at their frontiers. They held Red Tower, Torzburg, Vulcan, and Ghimes Passes. Von Falkenhayn had to be satisfied with ridding Transylvania of the Rumanians as well as capturing the majority of their equipment.

By November 1, the Ninth Army stood ready to mount the counterinvasion; however, small contingents of Rumanians on the passes and snows were sufficient to stall them for the time being. Von Falkenhayn had learned to be cautious with his resources. Ludendorff sensed this as procrastination but also believed that Rumanian forces had to be reduced further before attacking in such treacherous terrain. Instead of pressing the Ninth Army into action, Ludendorff ordered von Mackensen to stop the Dobrudja advance and cross the Danube with the intention of capturing Bucharest.[55]

Von Mackensen's battalions were already meeting heavy resistance in the Dobrudja. Along with the Rumanian divisions transferred from Transylvania, the Russians began sending additional reinforcements along with a new commanding general. The front seesawed for a number of days. These reverses were largely a result of thinning the Central Powers' lines to consolidate forces for crossing the Danube. Farsighted enough to understand what a devastating blow an attack on Bucharest would cause Rumanian forces, von Mackensen steadfastly pursued that objective rather than reinforce a faltering line.

On November 25, vanguards of the 217 German Infantry Division, the 1 and 12 Bulgarian Infantry Divisions, and the 26 Turkish Infantry Division crossed the Danube at Sistova under cover of a thick fog.[56] By using the islands there, the divisions completed the crossover in three days.[57] Defended only by the Rumanian 20 Cavalry, the area was quickly overrun.

The Danube crossing created panic among the Rumanians. Within hours after hearing the news, the government fled Bucharest for Jassy, leaving the capital to the army. The army was equally in a panic. The reserves left to defend the capital had been squandered in the abortive invasion of Bulgaria. Their only course of action, which would give them sufficient soldiers to defend Bucharest, was to pull the units off the passes and away from the Dobrudja.

Reinforced with five divisions, the Central Powers' Ninth Army renewed their part of the counterinvasion. Diversionary attacks were launched against the eastern passes in the first days of November with the hope of decreasing emphasis on the Vulcan Pass. This pass provided both tactical and strategic opportunities. Tactically, it was short, wide, and lightly defended. Strategically, the pass was within 50 miles of the railhead at Tragul-Jui in Rumania. On November 10, the German 301 Infantry Division forced Vulcan Pass and within five days secured the railhead.

The remnants of the Rumanian First and Second Armies offered weak but effective rear guard actions as they withdrew from the eastern passes. By November 26, the majority of the Wallachian plain was under Central Powers' control. With his right flank at Resi De Vede, von Falkenhayn was drawing near a link-up with von Mackensen's polyglot force that was advancing toward Bucharest. Hoping to slow down the advance, Rumanian forces attempted to drive a wedge between the two German commanders' forces at the Arges River southwest of Bucharest. Fighting lasted for three days. For a time, a German division was encircled and isolated, but at the last moment, the trap was broken by a Turkish division.[58] The passion of the attack didn't hold and the Rumanians began to withdraw again. As they relinquished the land, the soldiers began firing the countryside. Polesti's oil fields were set on fire in an effort to deny their exploitation but the Germans did not stop.

By December 4, von Mackensen's force had combined with the Ninth Army[59] and was within 12 miles of the capital; close enough to begin artillery barrages as well as aerial bombardment from airplanes and zeppelins.

Von Mackensen probably expected the Rumanians to resist his capture of Bucharest. The city was ringed by fortresses not unlike Namur and Liege. Like those two cities, the Bucharest forts had been built with nineteenth-century technology. A twentieth-century cannonade would have reduced them to rubble. Considering this and the fact that an estimated 120,000 soldiers[60] were needed to adequately man the forts, General Averescu, defender of the capital in his capacity as Third Army commander, decided to evacuate the capital and fall back to a line on the Polesti-Bukzen road just to the north. On December 5, ten days after crossing the Danube, Field Marshal von Mackensen entered Bucharest in his prized Hussar's uniform astride a snow white horse. He was the very picture of the conquering medieval warlord arriving in the enemy's capital as a result of twentieth-century technology.

The Rumanian defense on the Polesti road, even though reinforced by Russian units, quickly crumbled when faced with the combined strength of the Central Powers' armies. An estimated 75,000 Rumanians were taken prisoner. The retreat that followed was reminiscent of the Russian withdrawal from Galicia in 1915 or the flight of the Serbians in that same year. The Ninth Army kept the pressure on, pushing the Rumanian-Russian force back but never attempting an encirclement.

Following the advice of the Russians and the French, the remaining Rumanians withdrew to a prepared line along the Sereth River in January 1917.

Able only to hold 20 miles of this front, the Rumanians relied on the Russians to maintain the defense. Bratianu's five-month excursion into war had cost the nation's army to lose 150,000 soldiers to various categories of casualties of which half were dead.[61] Another 150,000 were wandering the countryside in search of their prospective units.[62]

REVOLUTION, LAST ACTS, CAPITULATION

The dawning of the eventful year of 1917 found Rumania almost entirely occupied by the Central Powers. Like Belgium, only a small strip of land on the east bank of the Sereth River still remained in Rumanian hands as a result of foreign intervention. The entry of Rumania, although coveted by the *Entente*, had done little to turn the tide of the Central Powers' victories in the East. On the contrary, Rumania's entrance and collapse had extended the Russian front by more than 400 miles and laid the rich lands of the Ukraine and Bessarabia open to attack. Throughout the winter of 1916 and 1917, von Mackensen's forces had bided their time in dealing with the defeated army. The respite allowed the French and Russians to reorganize the Rumanians along more modern lines much as they had done with the Serbians in Corfu.

France sent approximately 500 officers to begin training and restructuring the Rumanian army in December 1916. Russia, in an effort to release more of the front to the Rumanians, repatriated Rumanian prisoners taken from Austro-Hungarian units into the starving ranks of its ally. Rearmament was also under way by way of Russia. Captured German and Austro-Hungarian small arms and cannons taken during the Brusilov offensive were reallocated to the Rumanians. Amid this retraining, defense, and rearming, the fervor of revolution gripped the Russians in February 1917.

The Russian army began to crumble. Many units abandoned the Sereth line or openly fraternized with the enemy. In Jassy, soldiers appeared in bands and flew the red flag of revolution, calling on Rumanians to follow their example.[63] A Soviet was established at Socola. The problem worsened and finally caused the Rumanians to isolate the Soviet. On the line, retrained Rumanians replaced Russians. By May, the Rumanian army was once again at a strength of 300,000 and ready to launch a reconquest of their homeland.

The offensive was to be launched in conjunction with the Russian offensive in Galicia but, because of many Russian units' unwillingness to continue the war, neither attack began as planned. Unsure of when to begin, the Rumanians launched their attack along the Sereth a full 19 days after the Russians' had begun. By that time the Central Powers were already on the counteroffensive and Russian troops were either surrendering or deserting. Interestingly enough, the Rumanians did score some initial victories.

On the first day of their operation, the Rumanians opened a hole in the Austro-Hungarian line that was approximately 20 kilometers wide and three kilometers deep. They took 4,500 prisoners and 90 guns.[64] The Austrians pulled back to their original national border after five days of almost continuous fighting; however, the Russian troops who were supporting the drive at Zloczow succumbed to the revolutionary spirit and retreated from the battle, leaving the Rumanians' north flank in the air. Once again, the Rumanians had to stop their advance and go over to the defensive. Hurriedly, the Rumanian First Army assumed the gap created by the deserting Russians to await the Central Powers' reply.

Von Mackensen's reply began on August 4. He had been amassing troops for an offensive and these had remained untouched by the initial assaults of July 23. In quick succession, the Central Powers advanced, taking Marasesti and Focsani by August 11. In a counterstroke, unaided by the Russians, the Rumanians captured 1,100 and momentarily stopped von Mackensen, but the field marshal was no stranger to temporary setbacks. Within hours of their success, the Rumanians were forced back to Soveja, and Jassy was threatened. It was only bad weather and a last ditch defense which halted the Central Powers' armies in the Ocna Mountains in September.

For the second time, the Rumanian army was defeated and huddled in trenches. This time they held the line with no help from a foreign government. But once again the Central Powers chose not to press the advantage. To encourage the Russians to come to the peace table, the entire Eastern Front lay quiet. What could have been a very useful time for the Rumanians to refit was not to be. Russian deserters were roaming the countryside behind the line, looting and marauding as they returned to Russia. A sizable portion of the defending force was taken off the line to chase these bands out of the country.[65]

The Bolshevik Revolution ended any support from the Russians. Rumanian soldiers stood alone on the line in December 1917.[66] Out of a reorganized army of 300,000, only 60,000 soldiers remained.[67] King Ferdinand reluctantly agreed to an armistice and eventual peace talks when the Bolsheviks agreed to the Brest-Litovsk Treaty. The end to hostilities finally came in May 1918.

CHAPTER FOUR
A Return to the Galician Front

The entry of Rumania on the side of the *Entente* in August 1916 shifted Eastern Front activities from Galicia and Bukovina to Transylvania and the Wallachian Plain for the remainder of that year. By January 1917, the Rumanian front was a stabilized extension of the Galician theater and both sides had suspended fighting for one of the coldest winters of the war. In Germany, the people endured the "Turnip Winter" which led to an increase in food riots and growing dissatisfaction with the management of the war. In February, the world's eyes shifted to Petrograd. The Romanovs' totalitarian government was replaced by a provisional democratic government whose demeanor toward the war was vague.

The cause of the February Revolution continues to be controversial. Considering Bethmann-Hollweg's agenda of tearing down the Russian Empire from within, it would seem that the Germans had something to do with it. However, General Hoffman, promoted to chief of staff for the entire Eastern Front by February 1917, blamed it on the *Entente* who, he reasoned, had seen Russia flagging in its resolve to pursue the war. For him the blame fell solely on the shoulders of England and France who wanted a government that would continue the war in the East.[1] To the *Entente*, who was to blame for the Revolution was irrelevant. What mattered was whether Russia would be able to mount a spring offensive in conjunction with the Western and Italian Fronts' offensives. They were unaware of how much the Russian army in the trenches had changed.

Within days of the Revolution, Soldiers' Councils were established in every unit from the corps to the company. Their charge was to promote the rights of the soldier and spread revolutionary ideals. Often, officer opposition to their forming met with violent reprisals. Eventually the initial dust of confusion cleared and the common soldier found himself at the horns of a dilemma. Who was in charge of his well-being, the officer or the council? The Soldiers' Council was often the stronger of the two and, for a time, it seemed to want to correct many of the wrongs that had existed during the preceding three years of war. Following the example of the new government that had changed the majority of corps and divisional commanders,[2] councils conjured the soldiers to replace battalion, company, and platoon commanders with others popularly elected by the soldiers. For the most part, these newly installed officers were fellow revolutionaries who had enlisted in the ranks or men easily controlled by the council.

Yet the Revolution was not the rejuvenating shot many had thought it would be. The distribution of food and supplies remained inadequate. The only item with improved distribution was revolutionary literature unabated by censorship. But even these tracts proved to be confusing. This was caused by the diversity of ideals within the network of councils. Although professing unity in their beliefs, councils varied in their political outlook. For instance, Bolsheviks might control the division's council while Social Revolutionaries would be steering the regimental level and so on down the line until one arrived at the platoon. Each political faction had a different agenda for their charges. Those who supported the new provisional government under Alexander Kerensky talked of the soldiers' honorable obligation to continue the struggle against the Central Powers. They reasoned that only with the enemy completely off Russian soil could there be talk of peace. More radical groups called for the immediate cessation of all fighting and the redistribution of the land. To a soldier, the end of fighting had more appeal than any obligation to continue to die, and to a peasant, newly conscripted from the farm, the distribution of land was a compelling force for desertion. General Knox recorded an incident in which 2,500 conscripts had started for the front as replacements. By the time they reached the distribution points of their corps, only 400 were left.[3] In those early days, the Soldiers' Councils were unsure of how far they could push their power. They had deposed some officers in favor of more popular subordinates; however, strong officers still held sway in many units.[4] Particularly active unit Soldiers' Councils were often isolated along with their entire unit and disarmed, purged and reconstituted.[5] To forestall any further disintegration of the army, the Provisional government officials opined that the army should be involved as quickly as possible in an offensive. Kerensky called for such an attack within weeks of the Revolution.[6]

Planning the offensive took on different aspects than most officers had experienced. Gone was some of the ineptitude that Stavka had displayed throughout the war. Replacing it were the Soldiers' Councils through which all plans had to be coordinated. General Lavr Kornilov, the new commander of the Galician Front, went into negotiations with the various councils.[7] The first objection raised was the place of the attack. The councils preferred to attack at a "convenient place" which the Carpathian Mountains were not. Eventually, Kornilov and the councils were able to agree that the offensive's objective would be to gain the western bank of the Zlota Lipa and then to wheel northwest in an attempt to take Lemberg. Once an agreement on the objective was reached, negotiations began to fix the date of the attack. The XLI Corps of the Seventh Army agreed to attack at daylight on June 29 while other parts of the Seventh agreed to attack on the following day.[8] But a new phenomenon occurred which Kornilov was unable to understand. Some regiments declared their neutrality in any further military operations. Lacking half a division's strength in some instances, Kornilov appealed to Kerensky for help.[9] Kerensky personally visited the front to harangue the soldiers. Persuasive as he was, some units still decided to remain neutral during the offensive. Kerensky was dismayed at the soldiers' attitudes. At one moment they would enthusiastically support his efforts to continue the war; however, they would just as avidly support the next speaker who called for neutrality and an immediate end to the war.

As negotiations went on, Russian soldiers began to fraternize with their enemies all along the fronts. In some instances, Russians crossed over to the German trenches to spend a delightful day eating sausage and sipping cognac. In other cases, Germans went over to the Russian positions to lead discussions about who started the war and why it was going on. The standard story was that the *Entente* and the new government were continuing the war and not the Central Powers. The Russian soldier was told about the *Kaiser's* proposal for peace in the previous year and how the *Entente* had turned down the offer. Considering the sea of confusion that the Revolution had placed the Russian soldier in, the Germans' story was more trustworthy than anything to come out of Petrograd. Fraternization often led to the discussion of upcoming plans. Since Soldiers' Councils were democratic organizations and all aspects of a unit's operation had to have a consensus to be adopted, most soldiers knew the particulars of the upcoming offensive. Russian newspapers even discussed the upcoming offensive. Even more obvious were the preparations for the attack. It seemed that lessons had to be relearned. Reconnaissance airplanes took pictures of pioneers building bridges across the Dniester River and of uncamouflaged ammunition dumps. The news eventually reached the right ears. General Hoffman began moving reinforcements into key positions in anticipation of the attack.[10]

THE LAST OFFENSIVE

The offensive began with a barrage on June 29. It seems ironic that this last offensive should begin with the Russians better prepared for a sustained advance than at any time during the war. Along a 65-kilometer front, the Provisional Government's theater commanders had massed 31 infantry divisions supported by 800 light cannons, 158 medium guns, and 370 heavy caliber. Enormous shell reserves fed this enormous concentration of artillery. The barrage continued through the next day, finally lifting to further targets on the second day.

The Seventh Army was the first to begin the infantry attack at about midday on July 1 along an 18-mile front. Within a few hours, the soldiers had secured the first three lines of Austro-Hungarian trenches, capturing 18,000. Such large numbers of prisoners could in part be attributed to the placement

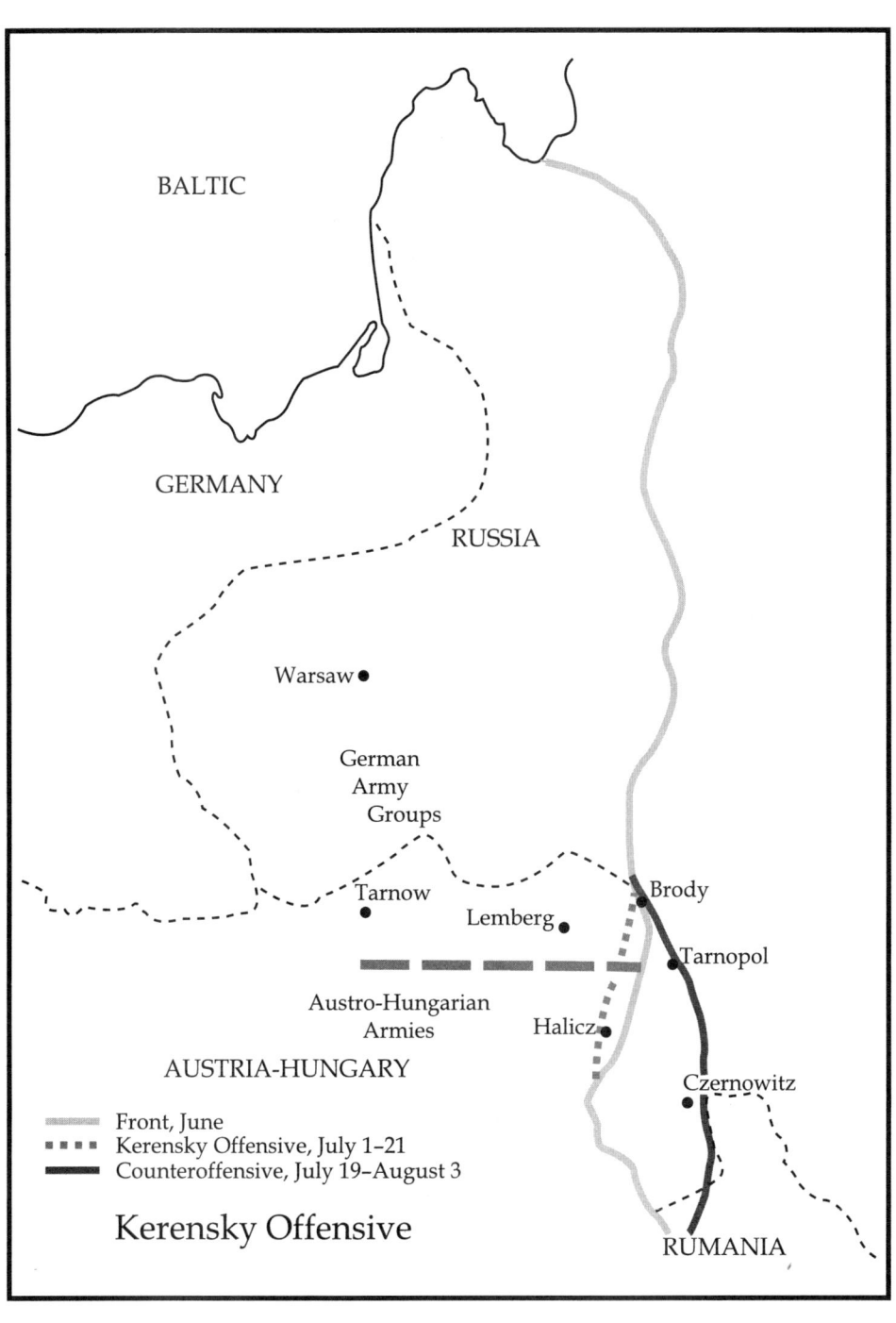

of Czech POWs in the trenches directly in front of the Austro-Hungarian 19 Division. When the Austrian Czechs heard their own language being spoken opposite them, they refused to fight and surrendered en masse.[11]

The Eleventh Army began its portion of the offensive on July 3 just north of the Tarnopol-Lemberg rail line. The Eighth joined in on July 8 along the Bistritza, pushing the front in by eight miles but not without difficulty. Originally, the Eighth was to begin their advance at three in the morning. The barrage lifted but the advance was not sounded. Instead, the soldiers argued about continuing the war. Delay upon delay was announced until finally, 12 hours later, the lead units climbed out of the trenches.[12]

General Hoffman recorded in his diary that the attack had begun and he hoped it would continue for the next 8 to 10 days.[13] What appeared to be gains by the Russians, advances of up to 20 miles, were really an elaborate scheme to overextend them and then encircle them. The only setback occurred because of a misunderstanding as to how the fresh troops that Hoffman moved into the area should be used. Some were used to replace front line troops who were then sent into rest rather than held as reserve. The result was that many of the fresh troops were captured and there were no reserves to replace them.[14]

Lemberg loomed on the horizon on July 19 when massed German artillery at Zloczow opened fire. Behind the bursting rounds came the German 1 and 2 Guard Infantry Divisions and the 5 and 6 Divisions. Within hours, the

In remembrance of Turkish support in the Ukraine, February 1918

Author's Collection

Russian army disintegrated into small groups who were in full flight back to their starting points and beyond. Rumors abounded that some nameless units on the flanks had decided to withdraw their support even before the counteroffensive had begun, thus causing others to retreat fearing encirclement. A communiqué dated July 13 claimed that nine army units had abandoned their positions and retreated. These units included the Izmailovskii, the Mlynovski, the Jaeger, the Moscow, the Grenadier, and the Finland Guard Regiments plus the 74, 113, and 153 Infantry Divisions.[15] These allegations were denied on July 31 by the accused regiments. The Sixth Guard Division stated that its entire force had been involved in the attack and as a consequence had almost been wiped out. Claiming to have faced 200 guns while it had only 16 to reply with, the division's 3,500 soldiers had advanced. Within a short time they experienced the loss of 95 officers and nearly 2,000 soldiers.[16] The Jaeger Regiment stated that it was never in the line at the time of the supposed retreat. In later years, Kerensky said that the allegations of desertion were probably a prefabrication by Kornilov in the hopes of reinstating the death penalty, a punishment that had been outlawed almost immediately after the success of the February Revolution. In any case, pressure by the advancing Germans and flooding due to torrential rains caused the Russians to leave their gained ground and relinquish even more Russian soil.

The Germans opened a 24-kilometer hole in the Russian line into which Hoffman poured the Austro-Hungarian Third and Seventh Armies along with the *Südarmee*. By July 21, 6,000 prisoners and 70 guns were taken.[17] Bukovina and Galicia were both cleared of Russians by the twenty-fifth of the month but, like in September 1915, the Central Powers' armies discontinued pursuit because supply lines became overextended. It was apparent to all that the Russian army was no longer a viable force. Recognizing this fact, Kerensky began overtures toward peace through Swedish contacts.[18] Before that could be negotiated, Russia was shaken by another, more radical, Revolution.

BOLSHEVIKS, BREST-LITOVSK NEGOTIATIONS

The November Revolution put Vladimir Lenin and the Bolsheviks in power in Petrograd. Although the Bolsheviks were the strongest force in Petrograd, what gained them support in the countryside and with the army was their promise to redistribute the land and immediately end the war. Within days of assuming power, the Bolsheviks proposed a general armistice to take effect right away. In keeping with their belief that they, as workers, were fighting for the oppressed masses in making the peace initiative, they asked that the armistice go into effect on all fronts.[19] This was Lenin's first attempt at idealism in forging a peace. He had reasoned that in light of the increased food riots occurring in Germany it was just a matter of time before a revolution began there. Any delineation of the terms on the part of the *Reich*'s representatives would add fuel to the growing unrest and clamor for peace that he had seen from his perch while in exile in Switzerland. When the German government declined to extend the cease fire to all fronts and the uprising did not come, the Russians asked that the Germans at least agree to freezing all Eastern Front forces in place, disallowing any movement to reinforce the Western Front. In a show of faith, Berlin consented to this term with a stipulation that units already in transient or under orders to move to the Western Front would be allowed to proceed.[20] The Russians conceded. The armistice began on December 15 with an expiration date of 28 days later. During that time, Lenin, remaining true to his conviction that revolution in Germany was just a matter of time, hoped that popular sentiment among German workers and soldiers for an overall peace would lead to a continuation of a people's revolution there. The time was not right for the revolution to ignite in other lands.

Russian and Central Powers' delegates met at Brest-Litovsk to discuss a peace treaty on December 22. The Germans, led by Foreign Secretary Richard von Kuhlman, came with the intention of carving up those lands occupied by non-Russians in keeping with the political objectives of 1914. Count Ottokar Czernin, Austria-Hungary's foreign secretary, came to make peace at any price to save the Habsburg Empire. The Russians, eventually led by Leon Trotsky, came with four demands: no punitive indemnities, no forcible annexations, the right of self-determination for all nationalities, and the restoration of Belgian, Serbian, Rumanian, and Montenegrin independence along with a withdrawal from northern France.[21] Both Trotsky and Lenin had designed these terms as a delaying tactic. They still awaited the revolution in Germany which would negate the need for a peace treaty.[22] Since the Russian proposal for peace encompassed the entire war, the Bolshevik representatives asked to suspend discussions until the rest of the *Entente* delegates were present. When neither delegates nor replies came from the West, the Russians asked that the site for negotiations be changed to Stockholm.[23] The Germans objected and the conference resumed on January 22, 1918. With both sides' terms out on the table, serious negotiations began.

The Central Powers naturally rejected the Russians' terms outright. Von Kuhlman very quickly pointed out that the peace conference would deal only with the end to hostilities with Russia. In light of no reply to the Russian offer to mediate the peace for the rest of the *Entente*, the Bolsheviks agreed. As to the term of forcible annexations and self-determination of all nationalities, Count Czernin pointed out that territory already occupied by the Central Powers could not be regarded as annexations—they were the spoils of war. He went on to say that self-determination was the right of governments and not of peoples.[24] Those quasi-independent governments that existed within occupied German territory, such as the Warsaw Congress, had stated already that they wanted to sever their ties with Russia. That was self-determination. Individual nationalities within those areas had no rights to sovereignty. The Russians countered by remaining adamant about the immediate evacuation of all occupied land so that plebiscites could be held in the areas. After only four days, the conference adjourned to consult each others' governments.

Von Kuhlman remained firm in his demands for the territories of Poland and the Baltic provinces. Public sentiment in Berlin agreed. From Vienna, Count

Czernin received instructions to make a peace regardless of what the Germans wanted. The Dual Monarchy was nearing the end of its stored food and it was in need of the vast amount of grains the Russians could supply.[25] In Petrograd, Trotsky conferred with Lenin. Both agreed that the Central Powers' terms were unacceptable. To tear away those areas currently under occupation would cause a grave loss of population and economic instability. But the Bolsheviks had promised peace to the Russians and there would be no working around the Central Powers' demands by threatening to resume the war. It was then that Trotsky, in keeping with revolutionary ideas, hit upon a solution. On February 10 he rose to address the delegates in Brest-Litovsk:

> We are removing our armies and our people from the war...We are going out of the war. We inform all peoples and their Governments of this fact. We are giving the order for a general demobilization of all our armies opposed at the present to the troops of Germany, Austria-Hungary, Turkey, and Bulgaria.[26]

In effect, the Russians would go home without admitting defeat or recognizing any settlement which ended the war. General Hoffman's reaction was one of shock. Such an action was unprecedented he said, but Trotsky was unmoved. Another deadlock was about to form; however, Hoffman and the *OHL* understood what to do when political negotiations no longer proved fruitful. Hoffman dictated an ultimatum to the Bolshevik government along with new, more stringent peace terms. Lenin had 48 hours to either accept the new peace dictates or the war would be resumed. The Germans were under no obligations to their people to come to an immediate settlement. When no reply came in the specified time, the Eighth Army began the advance north toward Petrograd.

Whether the Central Powers fell into Lenin's trap or they were successful in calling the Bolsheviks' bluff is a matter of interpretation. For Lenin, the resumption of hostilities relieved him of any criticism that might arise if he accepted unfavorable peace terms. By all indications, he had been forced into making a peace to avoid any more bloodshed among the Russian soldiers and workers. For the Central Powers, the resumption of the war enabled their troops to occupy all of the Baltic provinces. The few Russian soldiers still manning the lines outside of Riga offered little resistance to the advancing Germans. Within a few days, units were in Narva, less than 60 miles from Petrograd. Although Lenin had sued for an immediate peace less than 24 hours after the Eighth Army had begun its offensive, Hoffman had delayed any answer until his forces had taken the entire Baltic coast. Each side had achieved a strategic objective that would not have been available unless hostilities had resumed. The delegates signed the final treaty on March 3. Germany had been victorious in the East.

THE UKRAINE

While negotiations were under way with the Russians, the Central Powers were approached by delegates from the newly formed Republic of the Ukraine. Ukrainian Progressives had broken from their Russian counterparts almost immediately after the February Revolution and had formed a "government" at Kiev.[27] Calling themselves the Ukrainian Central Council, or Rada, they had managed to gain internal political recognition from the Kerensky government in July 1917. The Rada viewed itself more as a nationalistic party than a socialist one although it did begin work on the redistribution of land. In September, with the help of the Kerensky government, it had purged itself of all Bolshevik sympathizers.[28] Therefore, it was prudent for the Rada to claim complete independence from Petrograd when the Bolsheviks came to power.

Independence did not shield the Ukraine from reprisal or the civil war that racked Russia almost immediately after Lenin came to power. The Rada's tacit support of General Aleksei Kaledin's White counterrevolutionary movement in its south and its refusal to allow the Red Army to cross its borders in pursuit of the Whites caused the Bolshevik government to declare the Rada as reactionary. Mikhail Muravev had formed a government sympathetic to the Bolsheviks in opposition to the Rada at Kharkov in the eastern Ukraine. Muravev began a military campaign against the Rada in December 1917.[29] The Rada looked desperately for any allies who could offer support against the internal threat. The *Entente* agreed to recognize the Ukraine as a sovereign nation provided they continued to support the war effort. Such a stipulation was inane since the new republic was without an army or war materials. The Rada reasoned that the *Entente* was too far away to offer any tangible help other than the shipment of war material which must pass through Bolshevik-held territory. And, without soldiers to wield the equipment, a non-Bolshevik Ukraine would be lost. For a time, the Rada looked to the newly released Czech POWs who had formed an army with the intention of leaving Russia to fight with the Entente against Austria-Hungary. However, the Czech Legion had a tacit relationship with the Bolsheviks.[30] Eventually, the Rada hoped that Ukrainians, who had been formed into nationalistic units within the Russian army in 1915 and 1916, returning from the front could be banded together to form an army. To their chagrin, most of the Ukrainian veterans either joined the Bolsheviks or declared their neutrality in the struggle. Other soldiers took to looting the countryside which was largely unprotected.[31] In a desperate move, the Rada sent delegates to Brest-Litovsk to negotiate with the Central Powers. Initially not accepted as a sovereign nation, the Central Powers chose to recognize the Rada as a legitimate government when negotiations with Trotsky broke down in January 1918. However, recognition didn't stop Muravev's advance.

Muravev ravished the Ukraine almost unopposed. By January 19, his forces were in possession of Poltava. Nine days later, after moving over 60 miles, they were shelling Kiev in preparation for an assault. The attack on the capital began on February 4 and lasted five days. In many places, the fighting was from building to building. Eventually, it became apparent that the Rada could not hold out. The day before Kiev finally surrendered, the Rada fled to Jhitomir. On that same day, its delegates at Brest-Litovsk secured a peace treaty with the Central Powers. In exchange for vast stores of grain and future harvests, the Rada delegates were able to secure the Cholm district in the Congress of Poland and Krolan in eastern Galicia.[32] However, the fall of Kiev and the flight of the Rada would have seemed to nullify the treaty. Neither Germany nor Austria-Hungary

A Return to the Galician Front

were about to let the promise of food fall into the abyss of another Bolshevik republic.[33]

Upon the request of the Rada, Central Powers' forces began the invasion of the Ukraine 10 days after signing the peace treaty. Auspiciously, the invasion's goal was to restore the rightful government and order in the countryside. Under Field Marshal Hermann von Eichhorn, the primarily German army of 450,000[34] advanced along the railway from Galicia and out of occupied Rumania.[35] Alternately fighting Bolsheviks and the Czech Legion who were attempting to escape recapture, the Germans managed to liberate Kiev by March 2. The advance continued unabated through April. Although the announced goal was to reinstall the government, the Central Powers' hidden agenda was to secure all the black earth region, capture the coal field of the Donetz Basin, and neutralize the Russian Black Sea fleet.[36] The most important item was the capture of the grain stores and their shipment to Germany and Austria-Hungary. This desire for immediate shipments brought about a confrontation between the Rada and the liberators.

In March 1918, shortly after the return of the Rada to Kiev, General Hoffman began noticing what he termed as difficulties.[37] In keeping with their socialist views, the Rada had reappropriated the land to the peasants. However, the constant turnover of governments along with the still existent threat of the Bolsheviks in the east caused confusion among the peasants. This uncertainty led to a lackadaisical view toward planting the spring crops. The peasants reasoned that since it was unsure whether the land would remain in their hands, why should they plant?[38] And, if they did not plant, then releasing any stored grains was foolhardy. The peasants resisted German and Austro-Hungarian collection details, often with armed violence.[39] Ukrainian agriculture was in ruin and the promised grain shipments that the Central Powers expected were not forthcoming. Hoffman opined that unless the Rada delivered, "we shall have to look for another" government.[40] Events in April led to just that.

From the Ukrainians' point of view, there was an almost continuous battle over who was in charge of the Ukraine. The Central Powers had come to the Ukraine based on the Rada's request to establish law and order. Once the Rada had been reinstalled at Kiev, it would have appeared likely that the Ukrainians would take over the land and government. However, General Eichhorn saw things a little different. Since the Ukrainian government was not fully in control of the country's internal workings, he reasoned, the Central Powers would assist the Rada in living up to the stipulations of the peace treaty.[41] Not only did Eichhorn institute grain collection measures, he also introduced compulsory labor in the fields.[42] The Rada saw these measures as infringing on their national sovereignty and called for an immediate suspension of the provisions. Eichhorn responded by having German troops surround the place where the Rada met and arrested every representative. His rationale to Berlin was that the government was party to the establishment of an anti-German league.[43] To replace them, Eichhorn installed Pavel Skoropadsky to the ancient title of Hetman. Skoropadsky had been a career soldier in the Tsarist army. His positions had included aide-de-camp to the Tsar but above all he was the wealthiest man in the Ukraine and controlled vast estates.[44] Hoffman viewed the change in government as a popular movement whereas Ludendorff, in his memoirs, simply stated that "this government disappeared from the scene"[45] and Hetman Skoropadsky assumed control. But the Hetman had even less control than the Rada had had.

A nationalistic fervor had inspired the independence movement in the Ukraine. When Skoropadsky went to Berlin to personally thank the *Kaiser* for liberating his country, the people became incensed. Peasants, not associated with the Bolsheviks, took up arms to resist grain collection parties. The most noteworthy of these peasant groups was led by the anarchist Nestor Makhno. Employing guerrilla tactics, Makhno's groups attacked collection teams as well as garrisons. In September 1918, he managed to defeat a superior Austro-Hungarian force near the village of Dibrivki.[46]

In response, Eichhorn instituted a reign of terror in which mass shootings and hangings replaced negotiations and persuasions.[47] These reprisals had little effect on the peasants at first. By the middle of July 1918, the attacks had caused approximately 19,000 casualties among the Central Powers' forces. The crowning touch to their resistance came on July 30 when Irina Kakhovskaia and Boris Donskoi assassinated General von Eichhorn.[48] As with most political assassinations, the death of von Eichhorn did not stop the terror or the steady advance of German troops across the Ukraine.

While the Rada was being replaced, German troops continued their liberation of the Ukraine as well as the Black Sea ports. Odessa was taken with little resistance on March 12 and Kharkoff by the end of April. The entire Ukraine was secure from external Bolshevik threats and internal unrest by August. This was shown in August when Simon Petylura, the new leader of Ukrainians outside of the country, called for a general revolt. The peasants chose to ignore the appeal. *OHL* thought the area passive enough to replace active service regiments with elderly *Landwehr* units. By November, 15 *Landwehr* divisions remained in the Ukraine to perform policing duties.[49]

CHAPTER FIVE
Armistice

The first six months of 1918 saw what appeared to be a revitalized German army marching to certain victory over exhausted *Entente* forces. In the East, the Central Powers had defeated Serbia, Rumania, and Russia while in the south, Austro-Hungarian forces, supported by prestigious numbers of German soldiers, were routing the Italian armies. On the Western Front, the genius of Ludendorff's planning staff brought the Reich's army to the very gates of Paris again. Long-range shells burst on the Champs Elysées. Yet beneath the veneer of success was the reality of an army on the verge of collapse. The vast reserve of men was exhausted. Those who would have reached conscription age in 1916 had been called up during the months of August through November of 1915. The class of 1917 had donned their uniforms in January 1916. Still the ranks needed filling. The class of 1918 was conscripted in September 1916 only to be followed by the final conscription, the 1919 class, in the summer of 1917.[1] Not satisfied with the young recruits, *OHL* ordered the reassignment of all unmarried soldiers under 35 from units in the East to regiments in the West in September 1918.

The zeal of the young, even augmented with older soldiers, did not stop the inevitable. By September 1918, the army on the Western Front was in a slow retreat along the very route they had followed in 1914. Soldiers were captured rather than killed or wounded at a much higher rate than prisoners were taken between July and September.[2] It was apparent to Ludendorff and his staff that the army could go on no longer. In a frantic meeting with the Kaiser, Ludendorff shocked the emperor by asking that the government forward a request for an immediate armistice before foreign troops invaded Germany.

On the morning of November 11, 1918, German soldiers on all fronts were ordered to stand down and begin an immediate withdrawal from the firing line

63

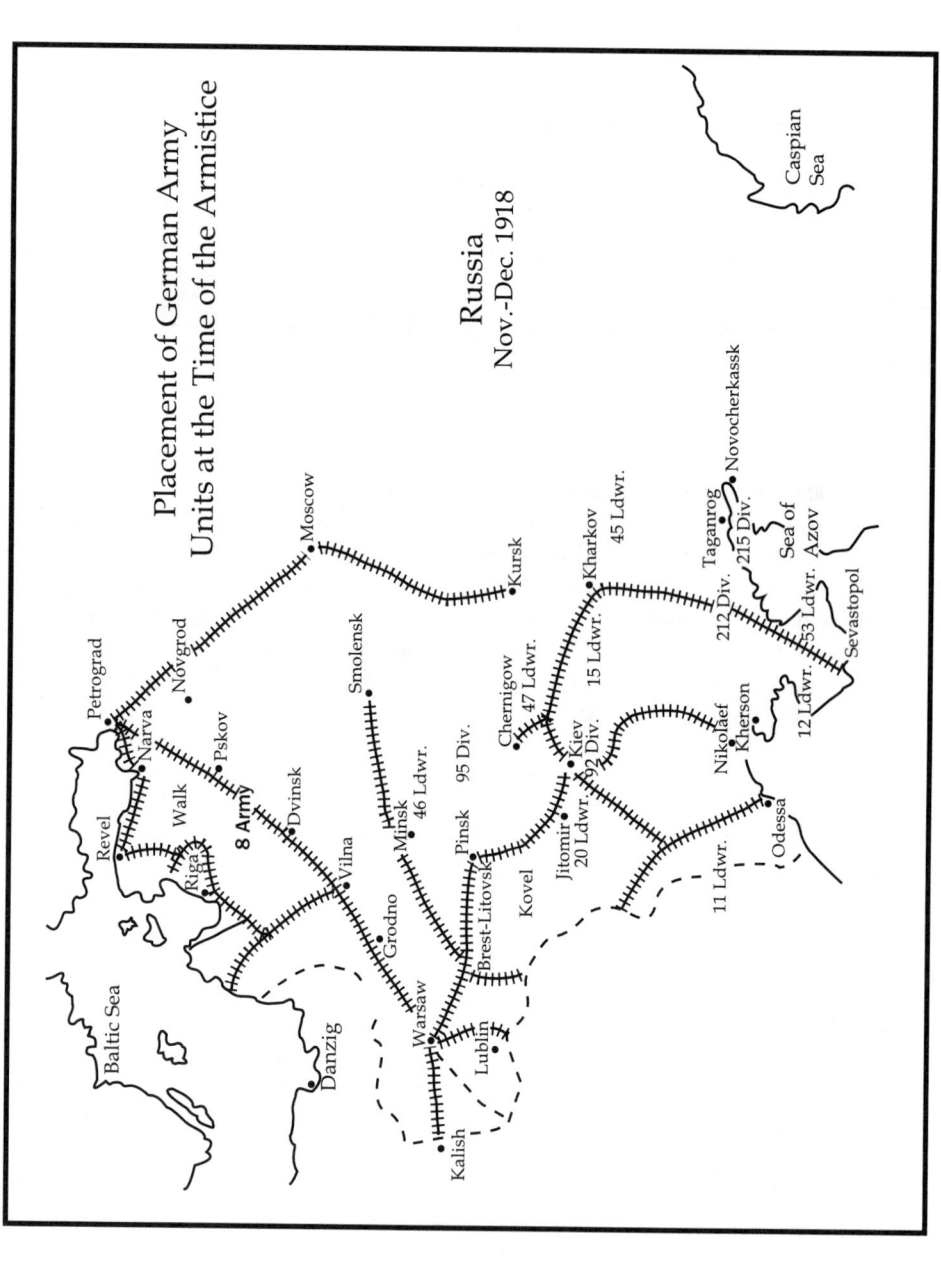

and all occupied areas. That evening, troop commanders read a proclamation to their men from Field Marshal von Hindenburg. It began by saying that the terms of the Armistice "obliged us to march back to the Homeland without delay"[3] and finished with an encouragement to each soldier to obey orders regarding the evacuation and stay together as a unit. Within minutes, enterprising soldiers began leaving for Germany on all fronts by any means available including the seizure of whole trains. Some individuals may have already been on the move east as early as November 9 when the news of the *Kaiser's* abdication reached the rear echelons. For the most part, units did stay together but not because of von Hindenburg's request.[4] Although the army had stood down, regulations still existed. Supply depots, despite their leanings toward supporting an immediate end to the war, continued to issue rations only to units and not to individuals. Coupled with this official attitude was the stabilizing factor of the *Soldatenraten* or Soldiers' Councils.

Soldatenraten sprang up in almost every unit with the introduction of universal suffrage. Modeled after Bolshevik soviets yet uniquely German, the councils reflected the apolitical stance of the majority of the soldiers. The main goal of each council was to secure their men's discharges as soon as possible. There were no wrongs to be righted as there had been in the Russian army. Reports of individual acts of tearing epaulettes from officers' tunics were received, and both officers and their men reacted with disgust. The majority of officers did not lose control of their commands as had been the outcome in the Russian army. In most cases, front line officers had endured the same deprivations as had their men; some even came from the same humble beginnings. This could be seen in who some of the soldiers elected to chair their unit council, their officers. It was not uncommon to find the company commander elected to the unit's council and then elected as a representative to the regimental or divisional council.

A few councils took it upon themselves to visit *OHL* at Spa to make revolutionary demands. These councils were dutifully listened to and then brought into the scheme of things regarding the Armistice terms. Many of the men were shocked and went away promising to do all they could to assist the evacuation. According to the terms of evacuating France and Belgium in 15 days, many units had to march 12 hours a day. General Hoffman, now commander of the Eastern theater, had similar meetings with *Soldatenraten*; however, the terms of the Armistice were different for the units under his command.

Under the terms of the cease-fire, German representatives had agreed to keep soldiers in the former Russian provinces in European Russia until such time as the newly developing republics were able to defend and police themselves. This provision came about for two reasons. The first was a realization that units in the Eastern theater were scattered over vast distances much of which was not serviced by rail. Prompt evacuation, even if there had not been a civil war going on all around, was impossible. But the second reason was of greater importance to the *Entente*. A quick withdrawal of forces from the East

Units at the Russian Front, Nov. 1918

Source: Histories of 251 Divisions of the German Army Which Participated in the War (1914-1918)

	UNIT	LAST REPORTED LOCATION	LAST DATE	HOME REGIMENT DEPOT
8th Army	2 Bavarian Landwehr	Livonia	May 1918	Bavaria
	3 Inf. Division	Courland		Pomerania
	14 Landwehr	On the Minsk-Smolensk Railroad at Orcha and Kockanovo		Posen/Silesia
	19 Landwehr	Estonia	Oct. 1918	Posen/Westphalia/Saxony
	23 Landwehr	Dvinsk area		Prussian Saxony
	29 Landwehr	Estonia	Sept. 1918	Rhine Province
	329 Infantry Rgt	Estonia	Sept. 1918	Posen/Lower Silesia
	205 Inf. Division	Narva	Sept. 1918	Prussian Saxony/Alsace
Ukraine	Bavarian Cavalry	Police force between Ukraine and Rumania		Bavaria
	4 Landwehr	East of Kiev	May 1918	6th Dist Silesia
	7 Landwehr	Rostov	July 1918	13th Dist. Wurtenberg
	15 Landwehr	Vardar Front	Sept. 1918	Brandenburg/Westphalia
	17 Landwehr	On the Don	Sept. 1918	Posen/Westphalia/Prussian Saxony
	18 Landwehr	Mohilev	Sept. 1918	Prussian Saxony/Mecklenburg
	20 Landwehr	Jitomir	Oct. 1918	Pomerania
	22 Landwehr	Stochod near Kiev	Sept. 1918	Pomerania/Baden
	35 Reserve	Ukraine at large	Oct. 1918	Saxony
	45 Landwehr	Kharkov	Oct. 1918	Saxony
	46 Landwehr	Minsk		Saxony
	47 Landwehr	Kiev		
	85 Landwehr	Polotsk	Oct. 1918	Lorraine/W. Prussia/Alsace
	93 Inf. Division	Kiev	Oct. 1918	Hesse/E. Prussia/Prussian Saxony
	95 Inf. Division	Gomel region	Oct. 1918	Prussian Saxony, Brandenburg
	212 Inf. Division	north of Kherson		Saxony
	215 Inf. Division	Sea of Azov		Prussian Saxony
Serbia	Alpine Corps	Near Nish	Oct. 1918	Bavaria and Prussia
	217 Inf. Division	Near Nish	Oct. 1918	Pomerania
	219 Inf. Division	Near Nish	Oct. 1918	Posen/Silesia
	302 Inf. Division	Macedonia (surrendered)	Oct. 1918	Posen/Westphalia/Saxony
Rumania	11th Landwehr	On Danube	Nov. 1918	Schleswig-Holstein/Mecklenburg
	16th Landwehr	Constanza	Oct. 1918	East Prussia/Brandenburg
	89 Inf. Division	Near Bucharest	Oct. 1918	Silesia/W. Prussia/Brandenburg
	92 Inf. Division	On the Danube	Oct. 1918	Westphalia/Thuringen
	218 Inf. Division		Oct. 1918	Westphalia/W. Prussia
	226 Inf. Division	near Bucharest		Pomerania

might bring the soldiers back before a formal peace treaty had been signed. These troops would greatly augment the western armies and may have led to a renewal of the war with a numerically superior German army firmly entrenched on the other side of the Rhine. It was to the *Entente*'s best interest to keep the army as scattered as possible. This fear of a resurgence of fighting could also be seen in the *Entente*'s refusal to release Central Powers' prisoners of war, but their adamant demands to have their own returned immediately. This attitude was further shown in the Armistice Commission's sanction of the internment in Hungary[5] of Germans returning from the Balkans and the disarming and internment of units crossing Polish demarcation lines.

Despite these terms, the majority of Hoffman's soldiers did not stay in place to defend the newly forming republics of Estonia, Latvia, Lithuania, the Ukraine, or those in the Caucasus from internal or external threats. Hoffman knew they would not and began arranging for the concentration of units and their shipment back to Germany.

In all the budding republics before the Armistice, the German army had been the stabilizing force which had purged the countryside of radical groups and staved off the Bolshevik forces. Whatever form the governments took, they would be necessarily hostile to the German presence especially in those areas where a coup had thrown out the more popular government. The Ukraine was such a place.

As many as four factions were operating in southern Russia in November 1918. Simon Petylura was in opposition to both the Hetman and the Rada he had succeeded. Petylura, with the backing of the Red Army, began to push Skoropadsky out of power almost as soon as the Armistice went into effect. Another faction was the anarchists led by Nestor Makhno. Makhno was friendly toward the Bolsheviks when the cause suited his followers. In opposition to the Hetman, his anarchists controlled a number of villages that adopted anarchist principles. The third faction were the White Russians led by Anton Denikin. With the exception of Skoropadsky, none of these elements wished the Germans to remain as a peacekeeping force.

Coupled to these antagonists inside the Ukraine, the new republics forming outside the area were just as adamant about getting rid of the Germans. The Poles, who had set up a government in Warsaw, were already establishing their national rights to parts of the former German and Russian empires. The rapidity with which they were able to establish a military presence on the Polish plain can be attributed to both the German and Russian governments. The Germans had disbanded and disarmed their I Polish Corps in June 1918, but rather than amalgamate the soldiers into other units, they had chosen to send them home. In contrast, the Kerensky government had simply allowed the Poles to return home fully armed. Vladimir Littauer, a Russian cavalry officer, recorded in his memoirs that the Poles mounted, fully equipped, as a unit one morning and left.[6]

The route for the evacuation of Russia was negotiated between General Hoffman, the Poles, and any other faction that might represent an authority

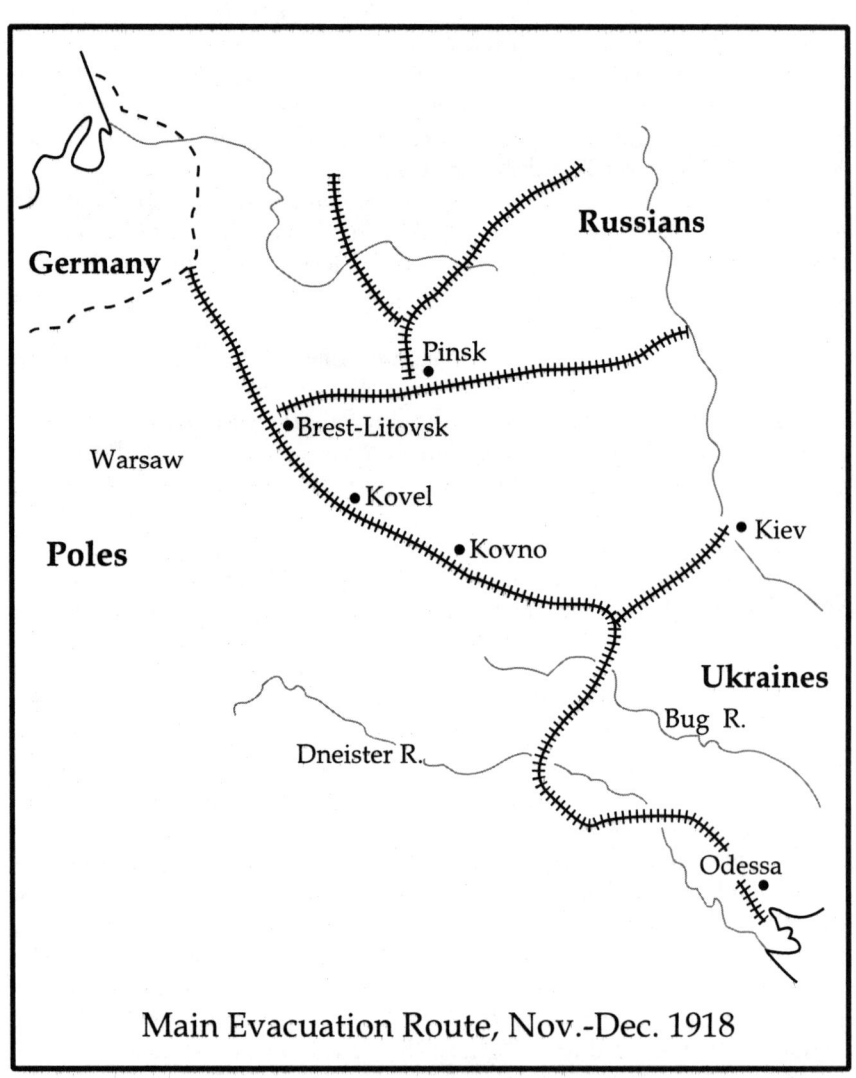

along the rail lines. Various massing points were established at Odessa, Kiev, and Pinsk. From these centers, trains took the soldiers across the border into East Prussia.

By late December most German units had left Russian soil. The exceptions were small groups in Estonia and Latvia who had stayed on to guard military stores or help hold back the Bolshevik army as it tried to retake former provinces and a contingent in the Ukraine at Nikolav on the shores of the Black Sea. This Black Sea location soon became a concentration area for Central Powers' soldiers and civilians. The *Entente* moved approximately 10,000 soldiers who had been in Palestine and Mesopotamia along with 10,000 civilians there in January 1919[7] to await transportation home.

Despite negotiations with both the Poles and Ukrainians, supply lines to the *Reich* were cut in early December when both the newly constituted governments stopped passage along its railways. Food and other necessities came from a weak French and English force which had occupied the Crimean peninsula in the previous month. Evacuation was finally made by ship through the Mediterranean and on to North Sea ports in May and April 1919, but as late as June, Germans soldiers were still at Nikolav. These last few men eventually made their way to Turkey where they were either interned or escorted to the border to make their own way back to Germany. Many did not return until 1920.

APPENDIX
Wenden, the Final Battle of the German Empire

Although the Armistice of November 11, 1918, ended the war on the Western Front, fighting, that involved German soldiers, on the Eastern Front, continued in two primary areas for the next year, along the Baltic coast and in Southern Russia. This continuation of hostilities came about as a result of a specific clause in the Armistice terms. Originally, the *Entente* negotiators had demanded the withdrawal of all German forces from occupied territory in both the East and West within 30 days of the Armistice's signing. However, the German delegation reminded the *Entente* representatives that Russia was a vast country torn by civil war. German troops in the Ukraine, southern and central Russia, and along the Baltic coast were in remote areas not necessarily serviced by a railroad. They opined that getting the soldiers together for evacuation could take a month in itself whereas transportation out of Russia would be dependent on either the Bolshevik government, whom the Germans had been fighting since November 1917, or the White Russians, who were of a mind to confiscate all military equipment, leaving the troops virtually defenseless as they crossed Bolshevik-held territories. Additionally, the German delegates continued, such fledgling republics as the Ukraine, Estonia, Lithuania, and Latvia were totally reliant on German divisions as a defense against the Red Army and for keeping internal order. To remove the army would mean that the new democracies might fall to Bolshevism thus negating President Woodrow Wilson's ideal of national self-determination in the former German, Austro-Hungarian, and Russian Empires. The *Entente* representatives acquiesced and rewrote the Armistice terms to state that German troops would remain in the portions of the former Russian Empire that they occupied to maintain order until such time as the Armistice Commission deemed that the new republics could stand on their own against internal and external threats.

Despite the good intentions of the German negotiators, the German army had already begun the process of withdrawing from their positions in Russia. Soldiers in Southern Russia and the Ukraine had commandeered all types of transportation to get them home before the Bolsheviks or the White Russians closed the rail lines that led to the Reich. German and Austro-Hungarian forces had maintained control of the railways throughout 1917 and 1918. Since the war was over, the Central Powers' soldiers could see no reason to defend anything outside of their own country. Those who were closer to the *Reich* naturally left first despite orders to the contrary. Their departure left a vacuum into which Bolshevik, White Russian, and the newly created Polish armies flowed. The German General Max Hoffman, chief of staff for the Eastern Front, recognizing that the soldiers would be on the move regardless of orders, had opened talks with all three factions vying for power in the Ukraine and on the Polish plain. The railroads were open but it was only a matter of time before they would be held ransom; payment to be military equipment.

Along the Baltic coast, the Eighth Army were so tantalizingly close to the *Reich* that many soldiers simply began walking. Behind them, Bolshevik sympathizers seized control. Farsighted individuals soon realized that the Bolsheviks not only aimed at regaining the Baltic coast, they also had Germany in their sights. Knowing that the men were unwilling to obey their officers, as was the custom in revolutionary times, some noncommissioned officers hit upon a scheme which would help the fledgling republics of the coast defend themselves and ensure Germans and their equipment were not molested unnecessarily in their evacuation. The plan was to ask for volunteers to guard military stores until they could either be destroyed or returned to the *Reich*. These new groups took on a status of *Freikorps*, a name not heard of since the early 1800s when disaffected Germans banded together in volunteer contingents to fight against Napoleon.

The small volunteer units entered their first battle near the Estonian-Russian border on November 22. A badly organized, poorly equipped but very numerous Red Army attacked Narva as an opening point for reclaiming the Baltic coast. Estonian nationalists had organized an army, numbering not more than a few thousand, from the Estonian regiments which the Tsarist army had formed in 1915 and 1916. Together, the Germans and Estonians held the Red Army for several days until they were outflanked by a Marine amphibious landing. Withdrawing slowly, the combined contingents fought delaying actions, but they were too small of a force to effect a victory or stalemate. Eventually, they were centered around Reval (Tallinn), their backs against the sea.

In Latvia, the story was approximately the same. Deeper in German-occupied areas, the middle class citizens of Latvia had declared their independence within days of the Armistice's signing. But a middle class–run republic was not representative of the majority of the people. There were two other population factions.

The largest portion of the population consisted of Letts. For many centuries, the Letts had been the unwilling workers of the vast German estates. These estates dated back to the Teutonic Knights and were run by the second, smallest part of the Baltic coast population, the Baltic barons. Even after the emancipation of the serfs in the 1860s, the Letts were not released from an obligation to till the land for landlords. Nikolai Lenin's government in 1917 appealed to the

Letts in that it promised a reappropriation of the land. The majority of the Letts threw their support behind the Bolsheviks. The Lettish regiments, formed in 1915 and 1916, became the most trusted military faction in the Red Army.

The second faction, the Baltic barons, found the Bolsheviks appalling and the middle class–run government not much better; however, an anti-Red government was preferable. The Baltic barons, coupled with the White Russians operating in the area, constituted a small army for defending the republic. Titled the *Landeswehr*,[1] it was commanded by a former German officer, Major Alfred Fletcher. *OHL* had ordered the major to volunteer his services to the Latvian government in December 1918.

In early December 1918, the Red Army's Latvian Light Infantry Regiments decided to go home. They advanced in three columns. The first column came out of Pskov with an objective of Riga while the other two groups marched on Liepaja from Zilupa, Polotsk, and Dvinsk. As they advanced across the countryside, they were greeted as heroes rather than an invading army. Support was everywhere among the Letts. As the soldiers approached a city or town, the people usually rose up against the defenders, overpowered them and delivered them to the regiments. The *Landeswehr* was pushed back just as the Estonians had been. By December 29, the Red Army was within artillery range of Riga. Prudently, Major Fletcher recommended to the republic's president, Karlis Ulmanis, that he evacuate the capital and take up position in Liepaja. Under the guns of a small British flotilla which had originally come to remove mines from the Gulf of Riga, the *Landeswehr* established a front that roughly corresponded to the Venta River, with Liepaja on the south and Ventapils on the north as anchor points.

In Estonia, backs against the sea, help came from the Finns in the way of volunteers and money and from the British in training and weapons. By February 1919, the Estonian army had pushed back the Red Army to its starting points of Iamburg in the north and Pskov in the south. The Finns withdrew almost immediately to leave the Estonians to themselves. The story was not the same in Latvia.

In early January, contingents of German volunteers, called *Freikorps*, began arriving at Liepaja. Ulmanis had signed a treaty with the new German government which promised Latvian citizenship to anyone who fought for Latvia against the Red Army. Former army and naval officers recruited combat-hardened veterans who had fought on either front and formed them into self-sufficient Stormtrooper groups. The leader of this *Freikorps* was another German officer whom *OHL* had volunteered, Major Joseph Bischoff.

Bischoff was a professional soldier. He had seen front line action for more than 12 years in Africa and on the Western Front. Methodically, he organized the volunteers as they arrived into what he termed to be the Iron Brigade. By the end of January the brigade included a full staff, three infantry regiments, trench mortar and machine gun companies, a cavalry regiment, a field artillery regiment, balloon and flamethrower sections, a field hospital, a company of

Comrades on the Fringe, Baltic *Freikorps*, April 12, 1919

Author's Collection

assorted armored vehicles, and airplanes.[2] Redesignated as the Iron Division on February 1, it was augmented by other *Freikorps* which had come en masse from Germany. These units were normally named after their commander: von Brandis, Eulenberg, Plehwe, Yorck. Others took their names from history: *Freiwilliger Jägerkorps*, and the *Eiserne Schar*. By March 2, *OHL* estimated that *Freikorps*' strength (termed the Sixth Reserve Corps) was about 10,900 men;[3] however, these figures were suspect since the Armistice Commission was pushing for a total demobilization of German forces. The figure was probably considerably higher. Contemporary writers placed the Iron Division's strength somewhere between 14,000 and 15,000 rifles while the *Landeswehr*, which was predominantly German, was approximately the same size.

During February, another volunteer arrived at Liepaja from the *Reich* to take over all supervision of the Sixth Reserve Corps, General Rudiger von der Goltz. Major Bischoff's organizational skills had paid off. Von der Goltz was able to call for an immediate offensive against the Red Army which was reported to be sending patrols into German territory around Memel.

The offensive began on March 3. Using the rail lines as the main source of transportation and an invasion route, the *Freikorps* took Jelgava on March 5. The Red Army retreated on Riga while the *Freikorps* occupied the rest of Latvia.

It was at this point that the intentions of Fletcher, von der Goltz, and Bischoff showed through their facade of fighting back the Bolsheviks, and the *Entente* powers became aware of a very potent threat to their victory in the war.

The Baltic barons had been biding their time regarding their representation in the Ulmanis government. In March they attempted to reopen talks with the president; however, Ulmanis would not give an inch. Von der Goltz proposed a compromise by which the barons would continue to furnish military support in return for a promise of a later discussion on governmental involvement. Ulmanis rejected the proposal. Seizing upon information he had received from an informant, Ulmanis arrested the barons who had been involved in an abortive cabal to take over Latvia the month before. Among those arrested was a *Freikorps* lieutenant. Indignant over the arrest of a comrade-in-arms, the von Pfeffer *Freikorps* rushed to the lieutenant's aid and arrested the entire Latvian officer corps of 550 men. These officers had been the one force that stood in the way of a baronial takeover. Taking advantage of the situation, Baron Heinrich von Manteuffel moved in to arrest Ulmanis, his ministers, and the parliament. Most of them were captured but Ulmanis managed to escape to the British flotilla.

The *Entente* reacted to the coup immediately. Sensing that von der Goltz was behind the takeover, they demanded his immediate recall, the restoration of Ulmanis, and the breakup of the von Pfeffer *Freikorps*. The German government refused to relieve the general, claiming that the uprising had been an internal matter. They went on to say that if von der Goltz had to return home then they would also take the entire *Freikorps* out of Latvia, leaving the *Entente* powers the responsibility of defending the coast against continued Bolshevik aggression. The *Entente* was not about to commit troops to another theater of intervention in Russia. They agreed that von der Goltz and the *Freikorps* could stay but any further action, such as an attack on Riga, must not occur. The German government agreed to the compromise; however, von der Goltz, in the capital awaiting the outcome of the ordeal, was furious but undaunted. Attempting to find a loophole, he asked if the requirement to remain inactive applied to Fletcher's *Landeswehr*. Although primarily made up of Germans, the *Landeswehr* was the national army of Latvia. Both the German government and the *Entente* powers agreed that meddling in Latvian internal affairs was not their concern. Von der Goltz immediately sent a telegram to Fletcher stating the restriction did not apply to his unit.

Fletcher's army began the attack on Riga at 10 o'clock in the evening on May 21. As the fighting progressed, *Freikorps* units, feeling obliged to support their comrades in arms, joined the battle. *Freikorps* units were already in place on Fletcher's right flank. These contingents consisted of *Freikorps* Yorck, von Brandis, Eulenberg, von Pfeffer, and Rieckoff. The combined force consisted of about 3,000 soldiers. By 12 o'clock on the following day, the *Landeswehr* and the *Freikorps* were in Riga. The city was cleared of Bolshevik forces by the following day. Red Army resistance was weak at best.

The *Entente* powers were appalled that their orders not to attack Riga had been ignored. They immediately imposed a naval blockade of the coast and sent a British commission to straighten out the matter. Meanwhile, the *Freikorps* and the barons launched a White Terror in Riga. Suspected Bolshevik sympathizers were hung or shot without trial. A curfew was imposed on the city. Anyone on the streets during the hours of six in the evening to six in the morning was shot without benefit of explanation or trial. Estimates of those executed in the White Terror were placed at 3,000.[4] The Letts had had enough of the *Freikorps*.

By mid-June, *Freikorps* units had reached Walk in the north and Daugavplis in the south. The Red Army had withdrawn from the Baltic coast and agreed to a tentative armistice which included recognition of the republics. For the Letts and the *Entente* powers, this meant that the *Freikorps* was no longer needed. To von der Goltz, the lack of a Red Army to fight meant that he could begin his plan to march on Petrograd. Accordingly, he ordered Fletcher and Bischoff to clear the remaining Baltic coastal area of any forces which would resist the march. Knowing that the Red Army was gone, von der Goltz was in effect directing his forces against the Estonian army and the newly organized Latvian army who were just as anti-Bolshevik as the *Freikorps*.

President Ulmanis had managed to evade arrest in March by escaping to the British flotilla. The British government had refused to recognize the baronial government that had taken his place. They had also decided that they must take action to end German domination in the area.

A shipment of 19,500 rifles, 500 machine guns, light artillery and ammunition arrived in Estonia along with applicable trainers.[5] The trainers brought together the former Latvian police force and soldiers who had escaped the *Freikorps'* arrest orders. By mid June, the new Latvian army consisted of two well-trained divisions. It was augmented by units from the Estonian army. The Estonian government viewed the advancing *Freikorps* as a threat to its national integrity. The total force consisted of more than 14,000 rifles.

Fletcher's *Landeswehr* encountered elements of the Estonian-Latvian force around June 3. Occupying Wenden (Cesis), he demanded that the force return to Estonian borders. It is evident that he did not recognize the Latvian contingent as any more than a part of the Estonian army. The ultimatum was refused and the combined armies attacked the *Landeswehr* on June 20.

Fletcher was in possession of Wenden. He was augmented by *Freikorps* von Jena, Malmedel, and Blöckelman. Bischoff's Iron Division was in position at Hinzenberg. Rittmeister Wilhelm Baron Engelhardt's Baltic Cavalry operated along the road just south of Wenden. The Estonian-Latvian units advanced along a broad front to engage. Outnumbered and outgunned, the *Landeswehr* was quickly put to rout. Engelhardt's cavalry advancing toward Wenden encountered soldiers retreating as individuals and throwing away their equipment in an effort to speed their flight.

Fletcher called for reinforcement from Bischoff who, instead of sending columns up to meet the retreating *Landeswehr* on the road, sent two *Freikorps* units along westerly roads. He undoubtedly hoped to outflank the enemy and pull up in its rear. Instead *Freikorps* Blankenburg and Kliest met stronger Estonian

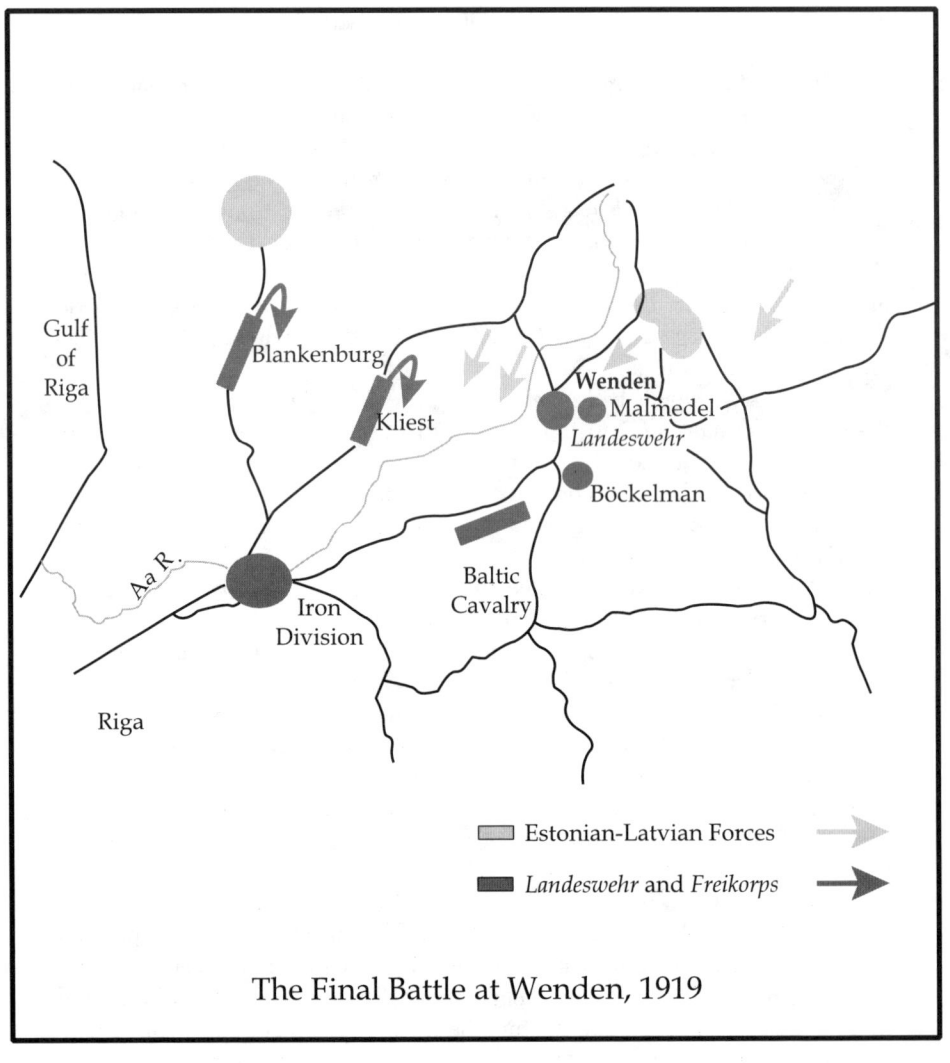

The Final Battle at Wenden, 1919

units which repulsed the flanking maneuver with ease. By midday on June 22, the *Freikorps* and the *Landeswehr* were hastily building entrenchments along the Dvina where they were able to hold.

Von der Goltz sued for an armistice on July 3, recognizing that further resistance would only cause undue losses. In keeping with the times, the Armistice terms were harsh. The Germans had to retreat to Jelgava (Mitau), abandoning Riga, reinstate Ulmanis as president, and agree to leave Latvia by August. Von der Goltz agreed to the terms.

The failure at Wenden, as with all non-successes, could never be attributed to any one organizational act. Fletcher did blame Bischoff for not sending in enough of a reinforcement at a crucial time. Von der Goltz cited Fletcher as the main reason, attributing his lack of control of the *Landeswehr* as a prime contributing factor. All three were adamant that exposure to Bolshevik ideals had sapped the spirit of the fighting men. Any or all of these may have been the reasons. It was evident, however, that some *Freikorps* members saw von der Goltz, Bischoff, and Fletcher as a threat to their continued personal existence. The *Freikorps* Baden had deserted to the Estonian-Latvian force and actually fought against their former comrades according to Ernst von Salomon, a participant in the battle.

When August came the *Freikorps* formed at the Jelgava rail station to board the trains which would take them back to Germany. A single figure walked onto the platform. Gaining the crowd's attention, Major Bischoff absolutely forbade any Iron Division member from leaving Latvia. The two months of inactivity had changed the soldiers. Instead of shouting him down, they cheered him and hoisted him onto their shoulders. That night, the soldiers held a torch light parade in his honor.

Over the next few days the *Freikorps* commanders met to discuss what they should do next. They agreed to join the White Russians and attempt a drive on Moscow via Smolensk. Reasoning that the *Entente* powers wanted the overthrow of the Bolsheviks at any price but the cost of their own soldiers, the leaders expected approval and funding of their venture once it was under way. Instead, they were once again rebuffed. In vengeance they attempted to capture Riga as a base of operations.

Their attack began on October 4 and continued until the twenty-first. Opposing them they not only found the Estonian-Latvian armies but also a combined British and French flotilla. Cut off from supplies by a blockade, the Germans soon succumbed to various maladies including starvation, dysentery, and decreasing ammunition supplies. Stalemate was finally achieved with the *Freikorps* in possession of the Dvina's left bank but little else.

On November 11, the Latvian army, aided by a bombardment from the Entente flotilla, began pushing back the Germans. Within four days, the Germans were back at Jelgava but the Latvians were not in the mood for an armistice. They continued to push and dog the *Freikorps* back to Germany. By November 30, all German units were out of the Baltic coastal republics and the dreams of a new empire were dashed.

NOTES

CHAPTER ONE

1. Jean-Louis Larcade, *Casques a Pointe Et Coiffures Prestigieuses de L'Armee Allemande, 1842–1918* (Paris: Jacques Grancher, 1983), pp. 157–60.
2. *The Times' History of the War*, vol. 1 (London, 1914), p. 227.
3. Bruce I. Gudmundson, *Stormtroop Tactics, Innovations in the German Army 1914–1918* (Westport, Conn. and London: Praeger, 1989), p. 19.
4. *Monarch Standard Atlas*, Revised Edition of 1906 (Chicago: Monarch Book Co., 1906), p. 171.
5. E. G. Ravenstein, *The New Census Physical, Pictorial, and Descriptive Atlas of the World* (Chicago: The Reilly and Britton Co., 1911).
6. Richard Bessel, *Germany After the First World War* (Oxford: Clarendon Press, 1993). In January 1915, there were 4,357,934 men on the active rolls of which 2,618,158 were in the field.
7. Ibid.
8. Fritz Nagel, *Fritz, The World War Memoir of a German Lieutenant* (Huntington, W.Va.: Blue Acorn Press, 1995), p. 26.
9. Under the Prussian Law of Siege of 1851, the military took over civil internal administration in August 1914.
10. Richard Bessel, *Germany After the First World War* (Oxford: Clarendon Press, 1993), p. 4.
11. *Monarch Standard Atlas*, Revised Edition for 1906 (Chicago: Monarch Book Co., 1906), p. 171. Seventeen-year-old men who were able to pass an examination on general subjects and agreed to equip and maintain themselves during the time of active duty were required to serve only one year of active service. From these ranks could come future NCOs and officers.
12. David Woodward, *Armies of the World, 1854–1914* (New York: G. P. Putnam's Sons, 1978), p. 30. Two years with the infantry or three years in the cavalry or horse artillery. Men seeking a military career could volunteer for 2-, 3-, or 4-year enlistments. These volunteers were normally selected for NCOs after signing up for their second term.
13. According to *The Times' History of the War*, military training for the *Ersatz* Reserve was almost entirely abolished in 1883.
14. Ian Drury and Gerry Embleton, *German Stormtrooper 1914–1918* (London: Reed International Books Ltd., 1995), p. 19.
15. John Ellis, *Eye Deep In Hell, Trench Warfare in World War I* (1976; reprint, Baltimore: The Johns Hopkins University Press, 1991), p. 36.
16. John Laffin, *Jackboot, The Story of the German Soldier* (New York: David and Charles Military Books, 1989), p. 126. Capt. Friedrich Schwerd, a medical officer, in collaboration with

Prof. August Bier proposed the idea of a steel helmet to *OHL*. Schwerd worked with Franz Marx on the design. Marx had done work in medieval armor restoration. The helmet's design may have been inspired by an engraving by Albrecht Dürer called "The Knight, Death, and the Devil" (1513). The knight wears a helmet called a *Schallern*.

17. John Ellis, *Eye Deep In Hell, Trench Warfare in World War I* (1976; reprint, Baltimore: The Johns Hopkins University Press, 1991), p. 125.
18. According to Fritz Nagel, in his memoirs, only the infantry was provided for in the cooking of the ration. Supporting artillery was given the ration and expected to cook it themselves in the 1914 army. Nagel, Fritz, *Fritz, The World War I Memoirs of a German Lieutenant* (Huntington, W.Va.: Blue Acorn Press, 1981, revised 1995), p. 37.
19. Ibid., p. 133.
20. This view of non-accuracy firing training proved to be a fallacy very early in the war. Erwin Rommel in his book entitled *Infantry Attacks* repeatedly mentioned that his unit often went into fire fights with the enemy less than 50 yards distant. At that close range, every shot must count.
21. Philip J. Haythornthwaite, *The World War One Source Book* (London: Arms and Armour Press, 1983), p. 67.
22. *The Times' History of the War*, vol.1 (London, 1914), p. 232.
23. *The Times' History of the War*, vol.1 (London, 1914), p. 242.
24. Bruce I. Gudmundson, *Stormtroop Tactics, Innovations in the German Army, 1914–1918* (New York and Westport, Conn.: Praeger, 1989), p. 8.
25. Ibid., p. 27.
26. Ibid., p. 35.
27. David Tschanz, "A Whiff of Death," *Command Magazine #33*, March–April 1995 (San Luis Obispo, Calif.), p. 46.
28. Victor Lefebur, *The Riddle of the Rhine* (New York: The Chemical Foundation Inc., 1927), pp. 32 and 33. The Hague Conventions stated that gases could not be used in projectiles. Since canisters were used, Germany theorized in the early part of the war, that they had not violated the Conventions.
29. Ian V. Hogg, "Bolimow and the First Gas Attack" from *Tanks and Weapons of World War I* (New York: Beekman House, 1973), pp. 17–19.
30. Bruce I. Gudmundson, *Stormtroop Tactics, Innovations in the German Army, 1914–1918* (New York and Westport, Conn.: Praeger, 1989), p. 38.
31. Philip J. Haythornthwaite, *The World War One Source Book* (London: Arms and Armour Press, 1983), p. 90.
32. *Tanks and Weapons of World War I* (New York: Beekman House, 1973), p. 21.
33. Alex Roland, "Greek Fire" from the *Quarterly Journal of Military History*, vol. 2, no. 3, spring 1990 (New York), pp. 16–19.
34. Michael Dewar, "The First Flame Attacks" from *Tanks and Weapons of World War I* (New York: Beekman House, 1973), pp. 48–50.
35. Ibid.
36. John Laffin, *Jackboot, The Story of the German Soldier* (New York: David and Charles Military Books, 1989), p. 129. Fritz Nagel, in his memoirs, stated that by September 15, 1914, out of the one million men deployed on the Western Front, two-thirds of the officers and two-fifths of the men were dead or wounded.
37. Walter Goerlitz, *History of the German General Staff, 1657–1945*, translated by Brian Battershaw (Boulder, Colo. and London: Westview Press, 1985), p. 169.
38. Bruce I. Gudmundson, *Stormtroop Tactics, Innovations in the German Army, 1914–1918* (New York and Westport, Conn.: Praeger, 1989), p. 30.
39. Ibid., p. 48. Krupp had designed a field piece for close artillery support but the cannon emitted too much smoke and was too large for moving through trenches. Rohr experimented with

captured Russian 76.2mm cannon and found them superior to Krupp's piece. They were modified by shortening the barrel and lightening the carriage; however, the shield stayed throughout the war.

40. Ibid., p. 49.
41. John J. Ellis, *Eye Deep in Hell, Trench Warfare in World War I* (1976; reprint, Baltimore, Md.: The Johns Hopkins University Press, 1991), p. 165.
42. Robert B. Asprey, *The German High Command at War, Hindenburg and Ludendorf Conduct World War I* (New York: William Morrow and Co., Inc., 1991), p. 154.
43. Max von Hoffman, *War Diaries and Other Papers*, vol. 1, translated by Eric Sutton (London: Martin Secker, 1929), p. 57.

CHAPTER TWO

1. Fritz Fischer, *Germany's Aims in the First World War* (New York: W. W. Norton Inc., 1967), p. 103.
2. Ibid., pp. 135 and 137. Before the war, the Russian census estimated that approximately 600,000 Germans were in the Ukraine. In the 1880s, they approached Bismarck asking him to support an independence movement in the area. Bismarck declined the invitation. Taras, Hunczka; *The Ukraine, 1917–1921: A Study in Revolution* (Harvard Union Press for the Harvard Ukraine Research Institute, 1977), p. 308.
3. Ibid.
4. Ibid., p. 274, The 1897 census placed the German population for the Baltic area at about three percent of the overall population.
5. F. R. Bridge, *The Habsburg Monarchy Among the Great Powers, 1815–1918* (New York: Berg Publishers, 1990), p. 345.
6. Girard Lindsley McEntee, *Military History of the World War* (New York: Charles Scribner's Sons, 1937), p. 10.
7. Brian Bond, *War and Society in Europe, 1870–1970* (New York: St. Martin's Press, 1983), p. 88. Fully eight-ninths of the German army was to attack France.
8. *The Times' History of the War*, vol. 1 (London, 1914), p. 211.
9. Girard Lindsley McEntee, *Military History of the World War* (New York: Charles Scribner's Sons, 1937), p. 107.
10. Ibid., p. 118.
11. Alfred W. F. Knox, *With the Russian Army, 1914–1917* (1921; reprint, New York: Arno Press and *The New York Times*, 1971), p. 434. One of the first divisions to be incorporated in the *Südarmee* was the 3rd Guard Division. They remained with the *Südarmee* throughout 1915 and the spring of 1916 when they were moved to the Western Front. Sustaining 57% casualties in the Somme in July, the division was pulled out for rest; however, in Sept., they were sent back to Galicia to participate in the counteroffensive at Halicz. Once again sustaining huge losses, they were sent back to the Western Front in November where they remained until the war's end.
12. Jim Bloom, "Max Hoffman, Phantom Genius of Germany's Eastern Front, 1914–1918," *Command Magazine* (Nov.–Dec. 1993, San Luis Obispo, Calif.). Hoffman was on the Eighth Army's battle staff at the beginning of the war. It was he who planned the Tannenberg and the Masurian Lakes battles. Surprisingly, he was not born into an aristocratic family; however, he managed to obtain a commission. Overweight and disliking exercise of any type, Hoffman did not fit the stereotypical image of the German officer. His understanding of the Russian military mind came from his tours of duty as an observer to the Russian army from 1898 to 1899 and as an observer to the Japanese army during the Russo-Japanese War.
13. Max von Hoffman, *War Diaries and Other Papers*, vol. 1 (New York: Harper Brothers Publishers, 1921), p. 53.
14. Ibid.

15. F. R. Bridge, *The Habsburg Monarchy Among the Great Powers, 1815–1918* (New York: Berg Publishers, 1990), p. 346.
16. Walter Goerlitz, *History of the German General Staff, 1657–1945*, translated by Brian Battershaw (Boulder, Colo. and London: Westview Press, 1985), p. 169. Von Falkenhayn was able to create 18–20 new divisions by taking one regiment out of each division. This was done largely because of the ease in managing a three-regiment division in trench warfare. One regiment was in the front lines, another was in reserve, and the third placed on rest.
17. Girard Lindsley McEntee, *Military History of the World War* (New York: Charles Scribner's Sons, 1937), p. 135.
18. Ibid.
19. W. Bruce Lincoln, *Passage through Armageddon* (New York: Simon and Schuster, 1986), p. 125.
20. Girard Lindsley McEntee, *Military History of the World War* (New York: Charles Scribner's Sons, 1937), p. 135.
21. W. Bruce Lincoln, *Passage through Armageddon* (New York: Simon and Schuster, 1986), p. 127.
22. William R. Griffiths, ed., *The Great War* (Wayne, N.J.: Avery Publishing Group Inc., 1986), p. 56.
23. E. G. Ravenstein, *The New Census Physical, Pictorial, and Descriptive Atlas of the World* (Chicago: The Reilly and Britton Co., 1911), p. 6.
24. W. Bruce Lincoln, *Passage through Armageddon, The Russians in War and Revolution, 1914–1918* (New York: Simon and Schuster, 1986), p. 239.
25. Ibid.
26. Ibid.
27. Ibid.
28. Ibid.
29. Florence Farmborough, *With the Armies of the Tsar, A Nurse at the Russian Front, 1914–1918* (New York: Stein and Day, 1975), pp. 109 and 115.
30. Manufacturing increased by over a thousand per cent. By spring 1916, factories delivered more than a million rifles and a billion cartridges. It was reported that each Russian soldier was equipped with a rifle and 400 cartridges. In artillery, more than 5,000 new cannons were produced along with more than 32,000,000 shells. Grand Duke Alexeev told the French that the Russian army had 7,000 cannons available with reserves of 1,000 rounds per gun. Even machine gun production increased to 5,000 a year. This flurry in manufacturing was augmented by large shipments of arms and artillery shells from Russia's allies and neutral nations which arrived at Vladivostok and Arkangelsk. From the U.S., France, and Italy came Winchesters, Gras-Kropatschek, and Lebel rifles. Added to this material glut was the arrival of approximately 2,000,000 men from the provinces and cities. Ted S. Racier, "When Eagles Fight, The Eastern Front in World War I," *Command Magazine* (Nov.–Dec. 1993, issue 35, San Luis Obispo, Calif.), p. 31.
31. Francis Whitting Halsey, *The Literary Digest History of the World War*, vol. 10 (New York: Funk and Wagnalls), pp. 93–95.
32. Alfred W. F. Knox, *With the Russian Army, 1914–1917* (1921; reprint, New York: Arno Press and *The New York Times*, 1971), p. 460.
33. W. Bruce Lincoln, *Passage Through Armageddon, The Russians in War and Revolution, 1914–1918* (New York: Simon and Schuster, 1986), p. 239.
34. Ibid.
35. William R. Griffiths, ed., The West Point Series, *Atlas of the Great War* (Wayne, N.J.: Avery Publishing Group, Inc., 1986), p. 239.
36. Ibid., p. 248.
37. Alfred W. F. Knox, *With the Russian Army 1914–1917* (1921; reprint, New York: Arno Press and *The New York Times*, 1971), p. 493. Knox related a humorous story in which he described how

artillery was concealed in one instance. Two gunners were told to imitate a battery in fire to draw enemy observers and counterfire. One gunner ran around an area with a bag of dust from which he would periodically let fly a handful. The other man spit kerosene from his mouth which he ignited.

38. Walter Hubatsch, *Germany and the Central Powers in World War, 1914–1918* (Lawrence, Kans.: University of Kansas Publications, 1963), p. 72.
39. Edward Gleichen, *Chronology of the Great War* (London and Novato, Calif.: Greenhill Books, 1988), p. 38.
40. Alfred W. F. Knox, *With the Russian Army, 1914–1917* (1921; reprint, New York: Arno Press and *The New York Times*, 1971), p. 440.
41. W. Bruce Lincoln, *Passage through Armageddon, The Russians in War and Revolution, 1914–1918* (New York: Simon and Schuster, 1986), p. 248.
42. Ibid.
43. Ibid.
44. Ibid., p. 257.
45. Ted S. Racier, "When Eagles Fight, The Eastern Front in World War I," *Command Magazine* (Nov.–Dec. 1993, issue 35, San Luis Obispo, Calif.), p. 35, and W. Bruce Lincoln, *Passage Through Armageddon, The Russians In War and Revolution, 1914–1918* (New York: Simon and Schuster, 1986), p. 248.
46. Glenn Torrey, "Rumania Declares War," *The Marshall Cavendish Illustrated Encyclopedia of World War I*, vol. 5, Editor in Chief Peter Young (New York: Marshall Cavendish, 1984), pp. 1591–95.
47. Alfred W. F. Knox, *With the Russian Army, 1914–1917* (1921; reprint, New York: Arno Press and *The New York Times*, 1971), p. 493. From June through September, the Russian Seventh Army reported capturing 2,000 Turks. The Russians found the Turks to be more ferocious fighters than the Germans. Knox alleges that the Turks killed Russian prisoners because they had heard the Russians were killing Turkish prisoners. Max von Hoffman, *The War and Lost Opportunities*, translated by Kegan Paul; (London: Trench, Truber, and Co., Ltd., 1924), p. 150, The Turks were allowed only one train to move both divisions, a feat which required several trains. Initially, they were put under the *Südarmee*'s command.
48. Alfred W. F. Knox, *With the Russian Army, 1914–1917* (1921; reprint, New York: Arno Press and *The New York Times*, 1971), p. 446.
49. Ibid., p. 461.
50. Ibid.
51. Ibid.
52. Ibid.
53. Ibid., p. 469.
54. Haldane MacFall, *Germany at Bay* (New York: Grossett and Dunlap, Undated), p. 244.
55. Alfred W. F. Knox, *With the Russian Army, 1914–1917* (1921; reprint, New York: Arno Press and *The New York Times*, 1971), p. 457.
56. Nicholas V. Golovine, *The Russian Army in the World War* (New York: Yale University Press, 1931), pp. 90, 91, and 97.

CHAPTER THREE

1. Although Rumania owed its independence to the Russians, the people identified themselves with a Latin origin dating back to the Romans. The closest European power that had similar roots, the Rumanians reasoned, were the French. On the other hand, the Rumanian aristocracy aligned themselves with Germanic roots.
2. Only part of the Dobrudja was a bog while the lower portion was described as a desolate, treeless expanse. The strip of land was acquired at Bulgaria's expense in the Second Balkan War. Before 1912, it belonged to Turkey. Bulgaria claimed it after the First Balkan War. But, Bulgaria had a falling out with its allies, Serbia and Greece, over the division of Macedonia.

Serbia and Greece turned on Bulgaria. While occupied with those two armies and doing badly, Rumania came out of its neutrality and invaded Bulgaria from the north. Bulgaria had no choice but to sue for peace. In compensation for their entry, Rumania took the Dobrudja. Territory was added with very little bloodshed.

3. Glenn Torrey, "Rumania Declares War," *Marshall Cavendish Illustrated Encyclopedia of World War I*, vol. 5 (New York: Marshall Cavendish, 1984), pp. 1591–95.
4. Although the Rumanian population in Transylvania represented the majority, the Magyars also pointed to the area as a part of their "homeland." A group of Magyars known as the Zeklers resided along the then Rumanian border. To give this land over to the Rumanians would be an abandonment of a Magyar tradition.
5. Glenn Torrey, "Rumania Declares War," *Marshall Cavendish Illustrated Encyclopedia of World War I*, vol. 5 (New York: Marshall Cavendish, 1984), pp. 1591–95.
6. Ibid. Not only an ardent nationalist, Ferdinand was married to a descendent of Queen Victoria. This may have influenced his outlook.
7. Francis Whitting Halsey, *The Literary Digest History of the World War*, vol. 8 (New York: Funk and Wagnalls Co., 1919).
8. Glenn Torrey, "Rumania Declares War," *Marshall Cavendish Illustrated Encyclopedia of World War I*, vol. 5 (New York: Marshall Cavendish, 1984), pp. 1591–95.
9. Ibid.
10. Ibid.
11. Sherman D. Spector, *Rumania at the Peace Conference, A Study of Diplomacy of Ioan I. C. Bratianu* (New York: Bookman Associates, Inc., 1962), p. 30. Bratianu allowed the release of two million tons of grain to the Central Powers in early 1916. To the *Entente*, an unspecified amount of oil was sold.
12. Glenn Torrey, "Rumania Declares War," *Marshall Cavendish Illustrated Encyclopedia of World War I*, vol. 5 (New York: Marshall Cavendish, 1984), pp. 1591–95.
13. Ibid. The British had just launched the Somme offensive and the Italians were pushing in one of the many Isonzo battles. These offensives were designed to relieve pressure from the French at Verdun. In the Somme, the British suffered in excess of 60,000 casualties on the first day.
14. Alfred W. F. Knox, *With the Russian Army 1914–1917* (1921; reprint, New York: Arno Press and *The New York Times*, 1971), p. 501.
15. Trevor Nevitt Dupuy and Wlodzimierz Onacewicz, *Triumphs and Tragedies in the East, 1915–1917*, New York: Franklin Watts, Inc., 1967, p. 60. Munitions were sorely needed not only because of the fact that until 1914 Austria-Hungary had been the main supplier but because the national armory in Bucharest had blown up and burned in July. None of the research indicated any foul play in the incident although one can assume that the Central Powers may have had something to do with it. In any event, the Rumanian army had only enough small arms ammunition to last a month.
16. Sherman D. Spector, *Rumania at the Peace Conference, A Study of Diplomacy of Ioan I. C. Bratianu* (New York: Bookman Associates, Inc., 1962), p. 34.
17. Ibid., p. 35.
18. Ibid.
19. The Salonika offensive was scheduled to begin on August 20.
20. Italy had been at war with Austria-Hungary since May 23, 1915, but did not declare war on Germany until August 27, 1916. There was a German presence in Rome. Von Falkenhayn, in his book *The German General Staff and Its Decisions, 1914–1916*, on p. 291, stated that he was aware of Rumania's intentions by the end of July 1916.
21. Sherman D. Spector, *Rumania at the Paris Peace Conference, A Study of Diplomacy of Ioan I. C. Bratianu* (New York: Bookman Associates, Inc., 1962), p. 33. The Hungarian Diet refused to endorse this offer but was overruled by the empire's foreign office in Vienna.
22. Ibid., p. 34.
23. Ibid.

24. Haldane MacFall, *Germany at Bay* (New York: Grosset and Dunlap, Undated), p. 181. General Sarrail had commanded the Third Army at Verdun in August 1914. He was credited with saving the French army's right wing through disobeying Joffre's order to abandon Verdun. His defensive measures secured the entire sector. Probably for going on the defensive rather than the offensive as Joffre wanted, Sarrail was relieved of the Third Army's command. Through political connections, he managed to avoid forced retirement as was the fate for so many commanders in those early months if they were in disagreement with Joffre. Those same political connections secured him command of Salonika.
25. Ibid., p. 237. 100,000 Serbs refitted at Corfu and retrained arrived in May 1916.
26. Ibid., p. 179. If historians can blame von Falkenhayn for fighting limited actions with no clear concept of victory, then the *Entente* should bear a responsibility for not recognizing opportunities that might have paid high dividends in the Balkans. In mid-August 1914, the Greek government offered its entire armed forces to the *Entente*, but Britain declined, hoping not to antagonize Turkey. With Turkey's entrance, Russia recognized the Greek army's importance and asked for its assistance in an attack on Turkey. Greece agreed but asked that Russia ensure Bulgarian neutrality. Petrograd never answered. When the Central Powers attacked Serbia, Greece asked the *Entente* to land troops on its territory to fill the positions that its army would leave behind as it marched off to assist Serbia. Before final arrangements could be made, the *Entente* started landing troops at Salonika. To avoid diplomatic embarrassment, Primier Venizelos protested the landings citing the country's neutrality but hurriedly rushed off to parliament with a request to declare war on the Central Powers. Before parliament could act, pro-German King Constantine relieved Venizelos of his position and reaffirmed Greek neutrality. The *Entente* ignored Greece's political stance and continued the landing unopposed by the Greek army. It was reasoned that to oppose the landing would place Greece on the Central Powers' side.
27. Erich von Falkenhayn, *The German General Staff and Its Decisions, 1914–1916* (1920; reprint, London: Books for Libraries Press, 1971), p. 322.
28. Alan Palmer, *The Gardeners of Salonika* (New York: Simon and Schuster, 1965), p. 76.
29. Ibid., p. 78. Curiously, the Bulgarians did not advance against the British. Instead, they chose to dig in opposite them along the Struma.
30. Trevor Nevitt Dupuy and Wlodzimierz Onacewicz, *Triumphs and Tragedies in the East, 1915–1917* (New York: Franklin Watts, Inc., 1967), p. 47. The number of divisions from the Western Front appears to be a matter of contention. Dupuy said 15; MacFall (p. 244) said 16 infantry; Ludendorff (p. 332) quoted only three divisions; von Falkenhayn stated only that "some" divisions were sent but they were replacements for spent units; Knox (p. 478) was able to identify only three divisions from the Western Front which arrived between August 8 and 12. These were the German Carpathian Division, 195 Infantry Division, and the 1 Infantry Division. Von Falkenhayn and Ludendorff agree that the mass of German divisions came from the Eastern Front. Additionally, two Turkish divisions joined Germany in bolstering Austro-Hungarian units.
31. Archduke Karl's army, although under his command, had as chief of staff the German General von Seeckt. As in any army, the chief of staff is the person who makes the plans and carries them out. It may be said that Austria-Hungary no longer had a commanding presence in this theater of operations after July 1916.
32. Francis J. Reynolds, editor in chief, *The Story of the Great War, History of the European War from Official Sources*, vol. 6 (New York: P. F. Collier and Sons Co.), 1916.
33. Ibid.
34. Ibid.
35. Erich von Falkenhayn, *The German General Staff and Its Decisions, 1914–1916* (London: Books for Libraries Press, 1920, 1971), p. 291.
36. Ibid.
37. Paul von Hindenburg, *Out of My Life*, vol. 1 (New York: Harper Brother Publishers, 1921), p. 241. *OHL* met with Austria-Hungary's and Bulgaria's counterparts at Pless on July 28, 1916, to

plan for an invasion of Rumania. Bulgaria agreed to enter the hostilities in return of the Dobrudja and territory in Macedonia. Turkey joined the plan on August 5.

38. Robert B. Asprey, *The German High Command at War, Hindenburg and Ludendorf Conduct World War I* (New York: William Morrow and Co., Inc., 1991), p. 251.
39. Erich von Falkenhayn, *The German General Staff and Its Decisions, 1914–1916* (London: Books for Libraries Press, 1920, 1971), p. 320. Although Germany agreed to deploy up to five infantry divisions and one cavalry division when Rumania attacked, in the meantime, defense was up to Austria-Hungary. To meet the expected hordes, the Dual Monarchy mobilized the *gendarme*, excise officers, alarm troops, *Landsturm* and their reserves, mountain troops, and single infantry units. A large amount of miners were also conscripted.
40. Nikolaus Krivinyi, "Rumania at War," *The Marshall Cavendish Illustrated Encyclopedia of World War I*, vol. 5 (New York: Marshall Cavendish, 1984), pp. 1597–1600. The flotilla and bridging equipment were brought down the Danube before hostilities began and anchored in Bulgarian territory.
41. Girard Lindsley McEntee, *Military History of the World War* (New York: Charles Scribner's Sons, 1937), p. 313.
42. Francis Whitting Halsey, *The Literary Digest History of the World War*, vol. 8 (New York: Funk and Wagnalls Co., 1919), p. 335. The bridge was 1,000 yards long with eight-mile long viaducts. Built at a cost of $35,000,000 in 1895, it was an engineering marvel of the time.
43. According to most maps of the time, there was only one rail line from the Black Sea into Rumania. There were two rail lines out of Russia but they junctioned at Polesti before going on to Bucharest and the rest of Wallachia.
44. Nikolaus Krivinyi, "Rumania at War," *The Marshall Cavendish Illustrated Encyclopedia of World War I*, vol. 5 (New York: Marshall Cavendish, 1984), pp. 1597–1600.
45. John A. Buchan, *Bulgaria and Rumania, The Nations of To-Day* (New York: Houghton Mifflin Co., 1924), p. 125. The defending force was placed at 39,000. Rumanian dead totaled 3,570. Bulgarian losses were 7,950.
46. Alfred W. F. Knox, *With the Russian Army, 1914–1917* (1921; reprint, New York: Arno Press and *The New York Times*, 1971), p. 483. The Russians thought that the Bulgarians would not fight them but Alexeev made the mistake of including Serbians in the Russian contingent. This placed an old enemy in the ranks of the Rumanians.
47. John A. Buchan, *Bulgaria and Rumania, The Nations of To-Day* (New York: Houghton Mifflin Co., 1924), p. 270. This line corresponded to Roman Emperor Trajan's wall.
48. Bertram Benedict, *A History of the Great War*, vol. 2 (New York: Bureau of National Literature, Inc., 1919), p. 339.
49. Ibid.
50. Nikolaus Krivinyi, "Rumania at War," *The Marshall Cavendish Illustrated Encyclopedia of World War I*, vol. 5 (New York: Marshall Cavendish, 1984), pp. 1597–1600. This included the 53,000 soldiers which had been held in reserve around the capital for its defense.
51. Alistair Horne, "Field Marshal Erich von Falkenhayn," *The War Lords* (Boston and Toronto: Little, Brown and Co., 1976), pp. 109–21.
52. The Ninth Army consisted of the 1 Infantry Reserve Division, 76 Infantry Reserve Division, 187 Infantry Division, 89 Infantry Division, 51 Infantry Division and the *Alpenkorps*.
53. Nikolaus Krivinyi, "Rumania at War," *Marshall Cavendish Illustrated Encyclopedia of World War I*, vol. 5 (New York: Marshall Cavendish, 1984), pp. 1597–1600.
54. Francis Whitting Halsey, *The Literary Digest History of the World War*, vol. 8 (New York: Funk and Wagnalls Co., 1919), p. 343.
55. Erich Ludendorff, *Ludendorff's Own Story, August 1914 to November 1918*, vols. 1 and 2 (New York: Harper and Brothers Publishers, 1919), p. 349.
56. Alistair Horne, "Field Marshall Erich von Falkenhayn," *The War Lords* (Boston and Toronto: Little, Brown, and Co., 1976), pp. 109–21.

57. George H. Allen, *The Great War*, vol. 4 (Philadelphia: George Barrie's Sons, 1919), p. 309. The Russians crossed the Danube at the same place during the 1877–78 War. It took them 33 days to move their forces across.
58. Erich Ludendorff, *Ludendorff's Own Story, August 1914 to November 1918*, vols. 1 and 2 (New York: Harper and Brothers Publishers, 1919), p. 354.
59. Ibid., p. 353. There was never any doubt that von Mackensen would take overall command of the Central Powers' armies once he linked his force with von Falkenhayn's Ninth Army.
60. Bertram Benedict, *A History of the Great War*, vol. 2 (New York: Bureau of National Literature, Inc., 1919), p. 356.
61. Girard Lindsley McEntee, *Military History of the World War* (New York: Scribner's Sons, 1937), p. 318.
62. Ibid.
63. John Buchan, *Bulgaria and Rumania, the Nations of To-Day* (New York: Houghton Mifflin Co., 1924), p. 273.
64. Francis Whitting Halsey, *The Literary Digest of the World War*, vol. 8 (New York: Funk and Wagnalls Co., 1919), p. 362.
65. In pursuing Russian marauders, Rumanian soldiers crossed into Bessarabia. To protect Rumanians in that province, Ferdinand extended his protection into the area. In retaliation, Leninist Russia declared war on Rumania on January 24, 1918.
66. Francis Whitting Halsey, *The Literary Digest History of the World War*, vol. 8 (New York: Funk and Wagnalls Co., 1919), p. 362.
67. Bertram Benedict, *A History of the Great War*, vol. 2 (New York: Bureau of National Literature, Inc., 1919), p. 365.

CHAPTER FOUR

1. Max Hoffman, *The War and Lost Opportunities*, translated by Kegan Paul (London: Trench, Truber, and Co., Ltd., 1924), p. 172.
2. Alfred W. F. Knox, *With the Russian Army, 1914–1917* (1921; reprint, New York: Arno Press and *The New York Times*, 1971), p. 634. Kornilov replaced Brusilov as front commander on July 18; VI Corps command was changed twice between February and July; three division commanders were changed along with eight of twelve regimental commanders just on the Galician front.
3. Ibid., p. 635.
4. Vladimir Littauer, *Russian Hussar, A Story of the Imperial Cavalry 1911–1920* (Shippensburg, Pa.: White Mane Publishing Co., Inc., 1993), pp. 248–49. Littauer was a cavalry squadron commander. Faced with an agitator, Littauer hit him once in the face and told him to desert or he would kill him. The man was gone the next morning.
5. Richard Abraham, *Alexander Kerensky, The First Love of the Revolution* (New York: Columbia University Press, 1987), p. 218.
6. Prior to the Revolution, the *Entente* had agreed to launch a spring offensive on all fronts. France and England expected the Provisional Government to honor the promise made by the Tsarist representatives. In fact, Kerensky was already making overtures for a separate peace as the planning of the offensive went on.
7. Alfred W. F. Knox, *With the Russian Army, 1914–1917* (1921; reprint, New York: Arno Press and *The New York Times*, 1971), p. 630.
8. Ibid., p. 641.
9. Richard Abraham, *Alexander Kerensky, The First Love of the Revolution*, New York: Columbia University Press, 1987, p. 215; Max von Hoffman, *War Diaries and Other Papers*, vol. 1, translated by Eric Sutton (London: Martin Secker, 1929), p. 181.
10. Max von Hoffman, *The War and Lost Opportunities*, translated by Kegan Paul (London: Trench, Truber and Co., Ltd., 1924), p. 182.

11. Norman Stone, "The Kerensky Offensive," *The Marshall Cavendish Illustrated Encyclopedia of World War I*, vol. 8 (New York: Marshall Cavendish, 1984), p. 2452.
12. Maria Botchkareva, *Yashka, My Life as Peasant, Officer And Exile* (New York: Frederick Stokes Co., 1919), pp. 209–12. In the line was the Battalion of Death, a unit composed of about 300 women whose purpose was to shame their male counterparts into taking the offensive. After waiting almost a day for the entire regiment to begin, the Battalion of Death was joined by the officers of the regiment; the soldiers stayed behind. Attacking independently, the battalion took the first three lines of German trenches, but reinforcements never came and the battalion had to return to their starting point.
13. Max von Hoffman, *War Diaries and Other Papers*, vol. 1, translated by Eric Sutton (London: Martin Secker, 1929), p. 186.
14. Max von Hoffman, *The War of Lost Opportunities*, translated by Kegan Paul (London: Trench, Truber and Co., Ltd., 1924), p. 145. The Russians took approximately 20,000 prisoners in the first days of the offensive.
15. Robert Paul Browder and Alexander F. Kerensky, *The Russian Provisional Government 1917 Documents*, vol. 2 (Stanford, Calif.: Stanford University Press, 1961), pp. 971–73.
16. Ibid., pp. 975–76.
17. Max von Hoffman, *War Diaries and Other Papers*, vol. 1, translated by Eric Sutton (London: Martin Secker, 1929), p. 188.
18. Max von Hoffman, *The War and Lost Opportunities*, translated by Kegan Paul (London: Trench, Truber and Co., Ltd., 1924), p. 174.
19. Bertram Benedict, *A History of the Great War*, vol. 2 (New York: Bureau of National Literature, Inc., 1919), p. 692.
20. Girard Lindsley McEntee, *Military History of the World War* (New York: Charles Scribner's Sons, 1937), p. 425.
21. Bertram Benedict, *A History of the Great War*, vol. 2 (New York: Bureau of National Literature, Inc. 1919), p. 695.
22. Ronald W. Clark, *Lenin: A Biography* (New York: Harper and Row Publishers, 1988), p. 335.
23. Ibid., p. 331. Stockholm had been the site of the Party's Fourth Congress in 1906 and it had become the Bolshevik finance center during the war years. For Lenin, Stockholm was a safe haven in which he hoped to get away from the domination of any *OHL* representative. Brest-Litovsk was in German-occupied Russia.
24. F. R. Bridge, *The Habsburg Monarchy Among the Great Powers, 1815–1918* (New York: Berg Publishers, 1990), p. 365.
25. Ibid., by mid-January, Austria-Hungary had less than two months supply of grain.
26. Ronald W. Clark, *Lenin: A Biography* (New York: Harper and Row Publishers, 1988), p. 335.
27. Michael Hrushevsky, *A History of Ukraine* (New York: Archon Books, 1970), p. 521.
28. Ibid., pp. 528–30.
29. W. Bruce Lincoln, *Red Victory* (New York: Simon and Schuster, 1989), p. 306.
30. Jonathan Allen, "A Long Road Home, The Czech Legion In Russia," *Command Magazine*, Sept.–Oct. 1993, p. 39 (San Luis Obispo, Calif.). The Czech Legion actually opposed the advancing Austro-German forces. Near the railway junction of Bachmach, the Germans fought a four-day battle with them. The Legion suffered 600 casualties but managed to escape along the rail line.
31. Michael Hrushevsky, *A History of Ukraine* (New York: Archon Books, 1970), p. 535.
32. F. R. Bridge, *The Habsburg Monarchy Among the Great Powers, 1815–1918* (New York: Berg Publishers, 1990), p. 366.
33. Wheat was the principal crop in the Ukraine. Average annual yield before the war was 7,716,000 short tons while potato harvests yielded 11,023,000 tons. Revyuk, Emil, Ukraine and the Ukrainians (Washington, D.C.: Friends of Ukraine, 1920), p. 17.
34. Michael Hrushevsky, *A History of Ukraine* (New York: Archon Books, 1970), p. 548.

35. Rumanian peace talks were going on simultaneously. They had signed an armistice on December 10, 1917. In January, Lenin's government declared war on Rumania because that country's forces had crossed into Bessarabia to restore law and order. Russian soldiers had taken to looting and terrorizing the Rumanian population in the province.
36. Erich Ludendorff, *Ludendorff's Own Story, August 1914 to November 1918*, vol. 2 (New York: Harper and Brother Publishers, 1919), p. 259.
37. Max von Hoffman, *War Diaries and Other Papers*, vol. 1, translated by Eric Sutton (London: Martin Secker, 1929), p. 213.
38. Ibid., p. 209.
39. Orest Subelny, *Ukraine, A History* (Toronto, Buffalo, London: University of Toronto Press, 1985), p. 358.
40. Max von Hoffman, *War Diaries and Other Papers*, vol. 1, translated by Eric Sutton (London: Martin Secker, 1929), p. 213.
41. Erich Ludendorff, *Ludendorff's Own Story, August 1914 to November 1918*, vol. 2 (New York: Harper and Brothers Publishers, 1919), p. 259.
42. Fritz Fischer, *Germany's Aims in the First World War* (New York: W. W. Norton Inc., 1967), p. 540.
43. Max von Hoffman, *War Diaries and Other Papers*, vol. 1, translated by Eric Sutton (London: Martin Secker, 1929), p. 215.
44. Skoropadsky could not speak Ukrainian. Considering that the Ukrainian independence movement was founded on nationalism, this fact would not have endeared him to the people. Additionally, Skoropadsky was related by marriage to General Eichhorn.
45. Erich Ludendorff, *Ludendorff's own Story, August 1914 to November 1918*, vol. 2 (New York: Harper and Brother Publishers, 1919), p. 260.
46. Paul Avrich, ed., *The Anarchists in the Russian Revolution* (Ithaca, N.Y.: Cornell University Press, 1973), p. 23.
47. W. Bruce Lincoln, *Red Victory* (New York: Simon and Schuster, 1989), pp. 155–56.
48. Ibid., p. 310. OHL considered fining the Ukrainian government for German deaths. The charge was graduated beginning with a private at 50,000 rubles up to 200,000 rubles for a general.
49. War Department, Document No. 905, *Histories of 251 Divisions of the German Army Which Participated in the War (1914–1918)* (Washington, D.C.,1920).

CHAPTER FIVE

1. Ian Drury and Gerry Embleton, *German Stormtrooper 1914–1918* (London: Reed International Books Ltd., 1995), p. 20.
2. Edward Gleichen, *Chronology of the Great War*, pt. 3 (London and Novato, Calif.: Greenhill Books, 1988), p. 97.
3. Stanley Weintraub, *A Stillness Heard Round the World, The End of the Great War, November 1918* (New York: Oxford University Press, 1987), p. 381.
4. The *Reich's* demobilization office reported in December 1918 that out of the 3.2 million soldiers in the western army, 1 million had set out on their own. The first 500,000 had done so in the first month of the Armistice. Corum, James S., *The Roots of Blitzkrieg, Hans von Seeckt and German Military Reform* (Lawrence, Kans.: University of Kansas), p. 74.
5. *United States Army in the World War, 1917–1919, The Armistice Agreement and Related Documents*, Historical Division (Washington, D.C.: Dept. of the Army, 1948), p. 1005.
6. Vladimir Littauer, *Russian Hussar, A Story of the Imperial Cavalry, 1911–1920* (1965; reprint, Shippensburg, Pa.: White Mane Publishing, 1993), p. 247.
7. *United States Army in the World War, 1917–1919, The Armistice Agreement and Related Documents*, Historical Division (Washington, D.C.: Dept. of the Army, 1948), p. 403.

APPENDIX

1. John A. Buchan, *A History of the Great War*, vol. 4 (New York: Houghton Mifflin Co., 1922), pp. 453–58. White Russians comprised the majority of the force with 2 companies of infantry and a squadron of cavalry.
2. Charles L. Sullivan, "The 1919 German Campaign in the Baltic: The Final Phase," *The Baltic States in Peace and War, 1917–1945*, edited by V. Stanley Vardys and Romuald J. Misiunas (University Park and London: Pennsylvania State University Press, 1978), p. 110.
3. *United States Army in the World War, 1917–1919, The Armistice Agreement and Related Documents*, Historical Division (Washington, D.C.: Dept. of the Army, 1948), PICA, Spa American Sector, WAKO, from Gen. Hammerstein to Gen. Nudant, 12 Mar. 1919, No. 13490, stated that as of 1 March *Landeswehr* Corps consisted of 8,200, Composite Res. Corps had 8,900, Headquarters No. 52 had 1,000, and the VI Res. Corps consisted of 10,300.
4. Robert G. L. Waite, *Vanguard of Nazism, The Free Corps Movement in Postwar Germany, 1918–1923* (1952; reprint, Cambridge: W. W. Norton and Co., Inc., 1969), p. 118.
5. Ibid., p. 119.

BIBLIOGRAPHY

Abraham, Richard. *Alexander Kerensky, The First Love of the Revolution.* New York: Columbia University Press, 1987.

Allen, George H. *The Great War.* Vol. 4. Philadelphia: George Barrie's Sons, 1919.

Allen, Henry T. *The Rhineland Occupation.* Indianapolis, Ind.: The Bobbs-Merrill Co., 1927.

Anderson, Benedict. *Imagined Communities.* London and New York: Verso, 1992.

Asprey, Robert B. *The German High Command at War, Hindenburg and Ludendorf Conduct of World War I.* New York: William Morrow and Co., Inc., 1991.

Aston, George. *The Biography of the Late Marshal Foch.* New York: Macmillan Co., 1932.

Avrich, Paul, ed. *The Anarchists in the Russian Revolution.* Ithaca, N.Y.: Cornell University Press, 1973.

Benedict, Bertram. *A History of the Great War.* Vol. 2. New York: Bureau of National Literature, Inc., 1919.

Benns, F. Lee. *Europe Since 1914.* New York: F. S. Crofts, 1934.

Berger, Maurice. *Germany After the Armistice.* Translated by William L. McPherson. New York: G. P. Putnam's Sons, 1920.

Bessel, Richard. *Germany After the First World War.* Oxford: Clarendon Press, 1993.

Bilmanja, Alfred. *A History of Latvia.* Westport, Conn.: Greenwood Press, 1951.

Bloom, Jim. "Max Hoffman, Phantom Genius of Germany's Eastern Front, 1914–1918." *Command Magazine.* Nov.–Dec. 1993, Issue 35. San Luis Obispo, Calif.

Blucher, Evelyn. *An English Wife in Berlin.* New York: E. P. Dutton and Co., 1920.

Bibliography

Bond, Brian. *War and Society in Europe, 1870–1970*. New York: St. Martin's Press, 1983.

Botchkareva, Maria. *Yashka, My Life as Peasant, Officer And Exile*. New York: Frederick Stokes Co., 1919.

Bouton, S. Miles. *And the Kaiser Abdicates*. Yale University, 1920.

Bridge, F. R. *The Habsburg Monarchy Among the Great Powers, 1815–1918*. New York: Berg Publishers, 1990.

Browder, Robert Paul, and Alexander F. Kerensky. *The Russian Provisional Government 1917 Documents*. Vol. 2. Stanford, Calif.: Stanford University Press, 1961.

Buchan, John A. *A History of the Great War*. Vol. 4. New York: Houghton Mifflin Co., 1922.

———. *Bulgaria and Romania, The Nations of To-day*. New York: Houghton Mifflin Co., 1924.

Carsten, F. L. *The Reichswehr and Politics, 1918–1933*. Oxford at the Clarendon Press, 1966.

———. *War Against War, British and German Radical Movements in the First World War*. Berkeley and Los Angeles, Calif.: University of California Press, 1982.

Churchill, Winston S. *The World Crisis*. Vol. 5. New York: Charles Scribner's Sons, 1957.

Clark, Ronald W. *Lenin, A Biography*. New York: Harper and Row Publishers, 1988.

Coetzee, Marily Shevin. *The German Army League*. Popular Nationalism in Wilhelmine Germany. New York: Oxford University Press, 1990.

Corum, James S. *The Roots of Blitzkrieg*. Hans von Seeckt and German Military Reform. Lawrence, Kans.: University of Kansas, 1992.

Dawson, William Harbutt. *Germany Under the Treaty*. 1935. Reprint, New York: Books for Libraries Press, 1972.

Diehl, James M. *Paramilitary Politics in Weimar Germany*. Bloomington, Ind. and London: Indiana University Press, 1977.

Drury, Ian, and Gerry Embleton. *German Stormtrooper 1914–1918*. London: Reed International Books Ltd., 1995.

Dupuy, Trevor Nevitt, and Wlodzimierz Onacewicz. *Triumphs and Tragedies in the East, 1915–1917*. New York: Franklin Watts, Inc., 1967.

Ellis, John J. *Eye Deep in Hell, Trench Warfare in World War I*. 1976. Reprint, Baltimore: The Johns Hopkins University Press, 1991.

Evertt, Susan. *World War I, An Illustrated History*. New York: Exeter Books, 1980.

Von Falkenhayn, Erich. *The German General Staff and Its Decisions, 1914–1916*. 1920. Reprint, London: Books for Libraries Press, 1971.

———. *Der Feldzug der 9. Armee Gegen die Rumanen und Russen 1916–1917*. Berlin: Verlag E. G. Mittler & Sohn, 1921.

Farmborough, Florence. *With the Armies of the Tsar, A Nurse at the Russian Front, 1914–1918*. New York: Stein and Day, 1975.

Fischer, Fritz. *Germany's Aims in the First World War*. New York: W. W. Norton, Inc., 1967.

Fosten, D. S. V., R. J. Marrion, and G. A. Embleton. *The German Army, 1914–1918*. London: Osprey Publishing Ltd., 1992.

Friedlander, Henry. *The German Revolution of 1918*. New York and London: Garland Publishing Inc., 1992.

Gerard, James W. *My Four Years in Germany*. New York: Geo. H. Duran, 1917.

Gilbert, Martin. *Churchill, A Life*. New York: Henry Holt & Co., 1991.

———. *Atlas of World War I*. New York: Dorset Press, 1970.

Gleichen, Edward. *Chronology of the Great War*. London and Novato, Calif.: Greenhill Books, 1988.

Goerlitz, Walter. *History of the German General Staff, 1657–1945*. Translated by Brian Battershaw. London and Boulder, Colo.: Westview Press, 1985.

Golovine, Nicholas V. *The Russian Army in the World War*. New York: Yale University Press, 1931.

Griess, Thomas E., series editor. *Atlas for the Great War*. Wayne, N.J.: Avery Publishing Group, Inc., 1986.

Griffiths, William R. *The Great War*. The West Point Military Series. Wayne, N.J.: Avery Publishing Group Inc., 1986.

Gudmundson, Bruce I. *Stormtroop Tactics, Innovations in the German Army, 1914–1918*. New York and Westport, Conn.: Praeger, 1989.

Halsey, Francis Whitting. *The Literary Digest History of the World War*. Vols. 8 and 10. New York: Funk and Wagnalls Co., 1919.

Haythornthwaite, Philip J. *World War One: 1914*. London: Arms and Armour Press Ltd., 1989.

———. *World War One: 1917*. London: Arms and Armour Press Ltd., 1989.

———. *World War One: 1918*. London: Arms and Armour Press Ltd., 1989.

———. *The World War One Source Book*. London: Arms and Armour Press Ltd., 1992, reprint 1993.

Von Hindenburg, Paul. *Out of My Life*. Vol. 1. New York: Harper Brothers Publishers, 1921.

Von Hoffman, Max W. *The War and Lost Opportunities*. Translated by Kegan Paul. London: Trench, Truber and Co., Ltd., 1924.

———. *War Diaries and Other Papers*. Vol. 1. Translated by Eric Sutton. London: Martin Secker, 1929.

Hogg, Ian V. "Bolimow and the First Gas Attack" from *Tanks and Weapons of World War I*. New York: Beckman House, 1973.

———. *The Illustrated Encyclopedia of Firearms*. 1978. Reprint, Secaucus, N.J.: Wellfleet Press, 1992.

Hohenzollern, Wilhelm. *The Kaiser's Memoirs*. Translated by Thomas R. Ybarra. New York and London: Harper and Brothers, 1922.

Horne, Alistair. "Field Marshal Erich von Falkenhayn," *The War Lords*. Edited by Field Marshal Sir Michael Carver. New York: Little, Brown and Co., 1976.

Hrushevsky, Michael. *A History of Ukraine*. New York: Archon Books, 1970.

Hunczka, Taras. *The Ukraine 1917–1921, A Study in Revolution*. Harvard Union Press for the Ukrainian Research Institute, 1977.

Hubatsch, Walter. *Germany and the Central Powers in World War 1914–1918*. Lawrence, Kans.: University of Kansas Publications, 1963.

Information Department of the Royal Institute of International Affairs. *The Baltic States, A Survey of the Political and Economic Structure and the Foreign Relations of Estonia, Latvia, and Lithuania*. Westport, Conn.: Greenwood Press, 1970.

Jones, Nigel J. *Hitler's Heralds, The Story of the Freikorps, 1918–1923*. New York: Dorset Press, 1992.

Jukes, Geoffrey, "The Brusilov Offensive from Victory to Failure," *The Marshall Cavendish Illustrated Encyclopedia of World War I*, vol. 5, Editor in Chief Brigadier Peter Young. New York: Marshall Cavendish, 1984.

Knight-Patterson, W. M. *Germany from Defeat to Conquest*. London: George Allen and Unwin Ltd., 1945.

Knox, Alfred W. F., *With the Russian Army, 1914–1917*. 1921. Reprint, New York: Arno Press and *The New York Times*, 1971.

Koch, H. W., "The Freikorps," *The Marshall Cavendish Illustrated Encyclopedia of World War I*, vol. 10, Editor in Chief Brigadier Peter Young. New York: Marshall Cavendish, 1984.

———. *A History of Prussia*. New York: Dorset Press, 1978.

Krivinyi, Nikolaus. "Rumania At War,"*The Marshall Cavendish Illustrated Encyclopedia of World War I*, vol. 5, Editor in Chief Brigadier Peter Young. New York: Marshall Cavendish, 1984.

Laffin, John. *Jackboot, The Story of the German Soldier*. New York: David and Charles Military Books, 1989.

Larcade, Jean-Louis. *Casques a Pointe Et Coiffures Prestigieuses De L'Armee Allemande, 1842–1918*. Paris: Jacques Grancher, 1983.

Lefebure, Victor. *The Riddle of the Rhine*. New York: The Chemical Foundation Inc., 1927.

Lincoln, W. Bruce. *Passage through Armageddon, The Russians in War and Revolution, 1914–1918*. New York: Simon and Schuster, 1986.

———. *Red Victory*. New York: Simon and Schuster, 1989.

Littauer, Vladimir. *Russian Hussar, A Story of the Imperial Cavalry, 1911–1920*. 1965. Reprint, Shippensburg, Pa.: White Mane Publishing Co., Inc., 1993.

Ludendorf, Erich. *Ludendorf's Own Story, August 1914 to November 1918*. Vols. 1 and 2. New York: Harper and Brothers Publishers, 1919.

Lutz, Ralph Haswell. *The German Revolution, 1918–1919*. 1922. Reprint, New York: AMS Press Inc., 1968.

MacFall, Haldane. *Germany at Bay*. New York: Grosset and Dunlap, Undated, Estimated at 1917.

Magocsi, Paul Robert. *Ukraine: A Historical Atlas*. Toronto, Buffalo, London: University of Toronto Press, 1985.

Manchester, William. *The Arms of Krupp, 1587–1968*. Boston and Toronto: Little, Brown and Co., 1968.

Marshall, S. L. A. *The American Heritage of World War I*. New York: Dell Publishing, 1966.

McEntee, Girard Lindsley. *Military History of the World War*. New York: Charles Scribner's Sons, 1937.

McDonald, Lyn. *1914–1918, Voices and Images of the Great War*. London: Michael Joseph Ltd., 1989.

Monarch Standard Atlas. Revised Edition of 1906. Chicago: Monarch Book Co., 1906.

Moyer, Laurence. *Victory Must Be Ours*. New York: Hippocrene Books, 1995.

Nagel, Fritz. *Fritz, The World War I Memoir of a German Lieutenant*. Edited by Richard A. Baumgartner, Huntington, W.Va.: Blue Acorn Press, 1995.

Nettl, J. P. *Rosa Luxemburg* (Abridged Edition). New York: Schocken Books, 1969.

Von Oertzen, F. W. *Die Deutschen Freikorps, 1918–1923*. Munich: 1938.

Orlow, Dietrich. *Weimar Prussia, 1918–1925, The Unlikely Rock of Democracy*. University of Pittsburgh Press, 1986.

The Outlook, A Weekly Newspaper, Vol. 112. May–August 1916. New York: The Outlook Co., 1916.

Page, Stanley W. *The Formation of the Baltic States, A Study of the Effects of Great Power Politics upon the Emergence of Lithuania, Latvia, and Estonia*. New York: Howard Fertig, 1970.

Palmer, Alan. *The Gardeners of Salonika*. New York: Simon and Schuster, 1965.

Bibliography

Peukert, Detlev. J. K. *The Weimar Republic, The Crisis of Classical Modernity*. Translated by Richard Deveson. New York: Hill and Wang, 1987.

Pipes, Richard. *Russia Under the Old Regime*. New York: Charles Scribner's Sons, 1974.

Racier, Ted S. "When Eagles Fight, The Eastern Front in World War I," *Command Magazine*. November–December 1993, issue 35. San Luis Obispo, Calif.

Von Rauch, Georg. *The Baltic States, The Years of Independence, Estonia, Latvia, Lithuania. 1917–1940*. Berkeley and Los Angeles: University of California Press, 1974.

Ravenstein, E. G. *The New Census Physical, Pictorial and Descriptive Atlas of the World*. Chicago: The Reilly and Britton Co., 1911.

Revyuk, Emil. *Ukraine and the Ukrainians*. Washington D.C.: Friends of Ukraine, 1920.

Reynolds, Franics J., Editor. *The Story of the Great War, History of the European War from Official Sources*. Vol. 6. New York: P. F. Collier and Sons Co., 1916.

Rommel, Ervin. *Infantry Attacks*. Calif.: Presidio Press, 1990.

Von Salomon, Ernst. *Das Buch von Deutschen Freikorpskämpfer*. Berlin: Wilhelm Limpert-Verlag, 1938.

———. *The Outlaws*. Translated by Ian F. D. Morrow. London: Jonathan Cape, 1931.

Showalter, Dennis E. "The German Soldier of World War I: Myths and Realities," *A Weekend With the Great War*. Weingartner, Steven, Ed., Shippensburg, Pa.: Cantigny First Division Foundation and White Mane Publishing Co., Inc., 1995, 1996.

Smith, C. Jay, Jr. *Finland and Russian Revolution, 1917–1922*. Athens, Ga.: University of Georgia Press, 1958.

Spector, Sherman D. *Rumania at the Peace Conference, A Study of Diplomacy of Ioan I. C. Bratianu*. New York: Bookman Associates, Inc., 1962.

Stone, Norman. *The Eastern Front, 1914–1917*. New York: Scribner, 1975.

Stumpf, Richard. *War, Mutiny, and Revolution in the German Navy, The World War I Diary of Seaman Richard Stumpf*. Translated by Daniel Horn. New Brunswick, N.J.: Rutgers University Press, 1967.

Sturley, D. M. *A Short History of Russia*. New York: Harper Colophon Books, Harper and Row, 1964.

Subelny, Orest. *Ukraine, A History*. Toronto, Buffalo, London: University of Toronto Press, 1985.

Sullivan, Charles L. "The 1919 German Campaign in the Baltic: The Final Phase"; *The Baltic States in Peace and War, 1917–1945*. Edited by V. Stanley Vardys and Romuald J. Misiunas. University Park and London: Pennsylvania State University Press, 1978.

Sullivent, Robert S. *Soviet Politics and the Ukraine, 1917–1957*. New York: Columbia University Press, 1962.

Tanks and Weapons of World War I. New York: Beekman House, 1973.

Taylor, Edmund. *The Fall of Dynasties, The Collapse of the Old Order 1905–1922*. Garden City, N.Y.: Doubleday and Co., Inc., 1963.

Tschanz, David. "A Whiff of Death," *Command Magazine* #33, March–April 1995. San Luis Obispo, Calif.

The Times' History of the War. Vols. 1–3. London, 1914.

Toller, Ernst. *I Was A German*. Translated by Edward Crankshaw. 1933. Reprint, New York: Paragon House, 1991.

Torrey, Glenn. "Rumania Declares War," *The Marshall Cavendish Illustrated Encyclopedia of World War I*. Vol. 5. Editor in Chief Brigadier Peter Young. New York: Marshall Cavendish, 1984.

Troyat, Henri. *Daily Life in Russia Under the Last Tsar*. Translated by Malcolm Barnes, Stanford, Calif.: Stanford University Press, 1979.

United States Army in the World War, 1917–1919, The Armistice Agreement and Related Documents. Historical Division, Washington, D.C.: Department of the Army, 1948.

Vincent, C. Paul. *The Politics of Hunger, The Allied Blockade of Germany, 1915–1919*. Athens, Ohio: Ohio University Press, 1985.

Waite, Robert G. L. *Vanguard of Nazism, The Free Corps Movement in Postwar Germany 1918–1923*. 1952. Reprint, Cambridge: W. W. Norton and Co., Inc., 1969.

War Department, Document No. 905, *Histories of 251 Divisions of the German Army Which Participated in the War (1914–1918)*. Washington D.C.: War Department, 1920.

Weintraub, Stanley. *A Stillness Heard Round the World, The End of the Great War, November 1918*. New York: Oxford University Press, 1987.

Winter, J. M. *The Experience of World War I*. New York: Oxford University Press, 1989.

Woodward, David. *Armies of the World, 1854–1914*. New York: G. P. Putnam's Sons, 1978.

The World and Its Peoples, Rand McNally and Co. St. Louis, Mo.: The Thompson Publishing Co., 1906.

Wrangel, Alexis. *General Wrangel Russia's White Crusader*. New York: Hippocrene Books, Inc., 1987.

INDEX

A

Alexeev, Grand Duke Mikhail, 29
Alpenkorps. See German Army
Austro-Hungarian Army
  casualties in 1914, 19
  casualties in Brusilov offensive, 28, 30, 40
Austro-Hungarian Units
  2 Army, 27, 28, 30
  3 Army, 22, 31
  4 Army, 22, 27, 28, 40
  7 Army, 28, 40
Averescu, Gen. Alexandru, 44, 49

B

Baltic Barons, 71
Bethman-Hollweg, Chancellor Theobald von, 17
  Eastern Front war goals, 17–18
Bishoff, Josef, 72–75
Bratianu, Premier Ion, 34, 36, 37
Brest-Litovsk, Treaty of Negotiations, 57–59
Brusilov, Gen. Alexei, 25
Brusilov Offensive, 25–30
  casualties in, 30
  results of, 30–31
Bucharest, 41, 48, 49
Bulgaria, Counterinvasion of, 43
Bulgarian Army
  response to Rumanian invasion, 41
  at Salonika, 39

C

Carol I, 34
Cernovoda, Rumania, 44
Constanza, Rumania, 44
Czernin, Count Ottokar, 58

D

Dobrudja, Rumania, 36, 41, 43, 46, 48

E

Eichhorn, Gen. Hermann von, 61–62
Estonia, 70, 72
Evert, Gen. Nikolai, 26, 29

F

Falkenhayn, Gen. Erich von, 16, 40, 46, 48
Ferdinand, King, 35, 37
Flamethrowers, 12–13
  inclusion in Assault Units, 14
Fletcher, Maj. Alfred, 72, 74
Freikorps, 72, 74

G

Gas Warfare, 10–12
  in artillery shells, 11
  casualties from, 12
  classifications, 11–12
  first German usage, 10–11
  Hague Convention outlaws, 10
  second German usage, 11
Georgian League, 17–18
German Army
  Alpenkorps, 46
  conscription, 5, 63
  corps districts, 1
  evacuation from Russia, 71
  food rations, 6–7
  infantry armaments, 7
  infantry equipment, 6–7
  manning increases, 2
  organization, 1914, 1–2
  reorganization, 1916, 13

rifle training, 7
specially constructed units, 13
tactics, 7–9
German Units
  8 Army, 19
  9 Army, 16, 19, 48–49
  10 Army, 16
  11 Army, 16, 22
  6 Reserve Corps, 73
  10 Brigade, 14
  Special Assault Unit, 14–15
Ghimes Pass, 43, 48
Goltz, Gen. Rudiger von der, 73–74, 77
Gorlice, Austria-Hungary, 22
Greek Army, 39
Grenades, Hand, 9–10

H

Hermannstadt, Austria-Hungary, 43, 46
Hindenburg, Field Marshal Paul von, 16, 31, 40, 65
Hoffman, Gen. Max von, 16, 21, 29, 52, 54, 56, 61, 65, 67
Hotzendorf, Gen. Franz Conrad von, 19, 28–31

I

Industrial Production, Russian, 23
Iron Division, 73

J

Jassy, Rumania, 48
Jelgava, Latvia, 73, 77
Jhitomir, Ukrania, 60

K

Karl, Archduke, 40
Kavalla, Greece, 39
Kerensky, Alexander, 53–54
Kiev, Ukrania, 60
Knox, Gen. Alfred, 25–26, 30, 53
Kornilov, Gen. Lavr, 53–54
Kowel, Austria-Hungary, 26, 29–30
Kuhlman, Richard von, 58
Kuropatkin, Gen. Alexei, 26

L

Landeswehr, 72–74
Landwehr, 6
Landsturm, 5–6
Latvia, 70–71
Leagues
  Georgian, 17–18
  liberation of Ukraine, 17–18

Lenin, Vladimir, 57
Liepaja, Latvia, 72
Ludendorff, Q.M. Gen. Erich, 29, 40, 48

M

Machine Gun, Maxim, 7
  inclusion in Special Assault Unit, 14
Mackensen, Field Marshal August von, 21, 41, 43, 46, 49
Makhno, Nestor, 62
Manteuffel, Baron Heinrich von, 74
Masurian Lakes, Battle of, 19, 35
Muravev, Mikhail, 60

N

Narva, Estonia, 56
Nikolav, Russia, 69

P

Petliura, Simon, 67
Petrograd, 54, 59
Plan XVII, French, 8
Pskov, Russia, 72

R

Red Tower Pass, 41, 46, 48
Reval (Tallinn), Estonia, 71
Riga, Latvia, 73–74
Rohr, Capt. Ulrich, 14
Rumania
  Central Powers' offers, 35
  invasion of, 46–49
  neutrality, 34–35
  Non-Aggression Treaty, 35
  treaty with *Entente*, 36–37
Rumanian Army
  casualties, 49–50
  organization, 40
Rumanian Units
  1 Army, 41, 49
  2 Army, 41, 49
  3 Army, 43
  4 Army, 43
Russian Army
  casualties in Brusilov Offensive, 29–30
  equipment deficiencies, 23
  demoralization, 31
Russian Units
  3 Army, 19, 22, 28
  4 Army, 19
  5 Army, 19, 54
  8 Army, 19, 27–28, 30
  Guard Army, 29–30
Rustchuk, Rumania, 44

Index

S

Salonika Front, 37, 39–41
Sarrail, Gen. Maurice, 39
Schlieffen Plan, 8, 18
  view of Eastern Front, 18–19
Seeckt, Gen. Hans von, 21, 31
Sereth Front, 49–51
Silistria, Rumania, 43
Sistova, Rumania, 48
Skoropadsky, Hetman Pavel, 61–62
Soldatenraten, 65
Stosstruppen, 15
Südarmee, 16, 19, 27, 28, 30–31

T

Tannenberg, Battle of, 19, 35
Tarnow, Austria-Hungary, 22
Torzburg Pass, Rumania, 41
Transylvania, 34, 36, 41, 44, 48
  Austro-Hungarian defense in, 43
  Rumanian plan to invade, 41–42
Trench cannon, 14 n. 39
Trotsky, Leon, 58–59
Turkish Army
  in Galicia, 29
  in Rumania, 48
Turtukai, Rumania, 43–44

U

Ukraine, 59–62, 70
  Central Council (Rada), 59–61
  League for the Liberation of, 17–18
Ulmanis, Karlis, 72, 75

V

Ventaplis, Latvia, 72
Vulcan Pass, 43, 48–49

W

Wenden, 75
White Terror, 75